LESSONS LEARNED SERVICING BOILERS

Things you should know when maintaining low pressure steam and hydronic boilers

---

LESSONS LEARNED SERVICING BOILERS ver1.03

If you have any questions or comments, please send them to

Ray Wohlfarth
www.FireIceHeat.com
834 Kerry Hill Drive Pittsburgh, PA 15234
Tel 412-343-4110 Fax 412-343-4115
ray@fireiceheat.com

This book is an accumulation of ideas that I have seen and or used in my 30 plus years inside a boiler room. It is not a design manual. This does not take the place of the boiler manufacturers written instructions, engineering, or code issues that may be in force in your locale. Please follow the boiler manufacturers written instructions that are included in the boiler installation manual. This does not take the place of a properly designed system from an experienced designer. Thank you for choosing to purchase and read this book. If you find an error, please let me know so that I could change it for the next issue.

Dedication to the following:

My family, Sheila, Jon, Ryan, Abby, and Conor

My mentor and friend, Dan Holohan, for his support and guidance.

My expert proof readers Jon Wohlfarth, Ryan Wohlfarth

I would also like to acknowledge and thank the following individuals and organizations for helping me in my research for this book.

Field Controls

Ron Lukcic with Chemway RIP Ron!

HVAC & Industrial Controls in Charleston, WV

Joel Grabe for helping me while writing this book.

Neutrasafe Corp

Rel-Tek, Inc.

Rite Boiler

Triad Boiler

Erin Holohan Haskell for the cover design

All the great folks at TF Campbell in Pittsburgh for their expertise

You may notice that I will sometimes repeat the same examples in different sections of the book. It is not because I am forgetful, it is my way of helping the technician to quickly diagnose the problem when the customer is looking over the shoulder asking, "How much longer?"

Table of Contents

Intro

While speaking with various technicians in my 3 decades in the field, I have come to realize that boiler service is becoming a lost art. Many of the younger technicians are not familiar with the maintenance and operation of boilers and the majority of the training has been what was passed down from technician to technician. A boiler is a complex piece of equipment that requires preventive maintenance to operate safely and efficiently. It is the intent of this book to familiarize you with the different components on a boiler and help you to understand the maintenance of boilers. I like working inside boiler rooms as you are surrounded by such raw power and energy. Since it is common that you would be alone inside that mechanical jungle, you need to protect yourself from some of the inherent dangers inside the boiler room.

Natural Gas This book will feature natural gas boilers and burners as they are the ones that I am most familiar with. Although I have worked on many fuel oil burners, I would classify myself as a dummy compared to others in the industry such as fine folks at the National Association of Oil & Energy Service Professionals.

Boiler service requires patience and this was one of the most difficult parts for me while learning to be a boiler technician as patience was not one of my virtues. There are very few instantaneous results when performing boiler service. For example, when adjusting the fuel to air ratio of a burner, the effects of your change could take several minutes for the flue gases to travel through the boiler and report the new settings to the combustion analyzer. The same goes for adjusting the water feed valve for a hydronic system. It is better to go with small changes.

It is not what was installed, it was how it was installed. You can get long life and efficient operation from almost any boiler. I have found that two items affect the operation and life of a boiler; proper installation and regular maintenance. If these are done correctly, the owner should have a long lasting efficient heating system.

When is a Boiler not a Boiler? I know that question sounds almost Zen-like. I was asked to look at a boiler in a sewage treatment plant by a contractor friend. He informed me that it was not working properly. When I arrived, I saw the strangest boiler I had ever seen. It looked like an armored boiler that should be installed in an explosive environment. As I was diagnosing the boiler, I felt that it may be air bound. I suggested that we open a vent valve on top and the sewage worker said, "That boiler is full of crap." I thought he was expressing his feeling of anger at the boiler but he was just being truthful. The boiler I was working on was actually called a "Sludge Heater" and they piped raw sewage through it to be warmed. I turned and glared at my friend. He laughed and said, "Well, it is sorta a boiler."

Safety in the Boiler Room.

Look around Look around and see if there are obvious dangers. Take a mental picture of the layout of the room in case of a malfunction. For instance, if a boiler pressure relief valve opened, it could quickly fill the room with steam. If this happens, there would zero visibility. In addition, steam displaces oxygen which could lead to disorientation or even death. Another concern is that steam could enter the electrical panels which may cause an electrical short or fire.

No One Ever Goes Into a Boiler Room A municipality near me recently enacted a law that requires carbon monoxide detectors to be installed in every hotel room. The interesting part of the law is that carbon monoxide detectors are not required in the boiler room, which is where the carbon monoxide will form. I told the fire marshal that it would be less expensive to simply have a carbon monoxide detector in the boiler room and he said "No one ever goes into the boiler room." If you are working inside a boiler room, please let someone know so that they will check on you once in a while.

Exit Doors Make a mental note of the location of all exits in case you need a quick retreat. Is there anything in the path like ducts or pipes that could impede your escape?

Emergency Door Switch The emergency door switch is part of the ASME CSD1 code that many municapalities follow and is a great idea. It is a switch located at every exit from the boiler room and will shut off the power to all the boilers when pushed. Some are attached to signal lights and audible alarms that will alert you if they are pushed. If these are not installed in the buildings you service, they should be recommended to the owner.

Clearance / Cluttered? For many boiler rooms, it is the storage place for everything that the owner does not know what to do with. In many instance, it could be a powder keg just waiting to be ignited. I have been in rooms where toxic chemicals, flammable fluids, and nasty stuff are stacked inside. Some boiler rooms are used to store gas powered lawn equipment or pool chemicals such as chlorine. These chemicals may produce toxic gases when burned. Be very careful, my friend.

Each boiler requires a certain amount of clearance around it for service. The clearance is to allow combustion air for the burner and in case of flame rollout. Check the installation manual and local codes for how much clearance is required around the boilers. It is typically about 12-30". Is there a clear path in case of an accident? A friend of mine was in a boiler room and the boiler relief valve opened. He ran to the exit to shut down the boiler and tripped over a drain pipe injuring his leg. He was able to limp to the shutoff switch to shut down the boiler.

Open The Doors Normally, I like to simulate the actual conditions that the boiler experiences so I can effectively trouble shoot the boiler. Since the boiler operates with the doors closed, I will perform my service under those same conditions. I will not do that until I am sure the boiler room is a safe place to work. When I enter a boiler room, I like to open the doors to the outside until I can be sure that there are no toxic gases or fuel leaks.

Flame Rollout	Flue Gases Condensed
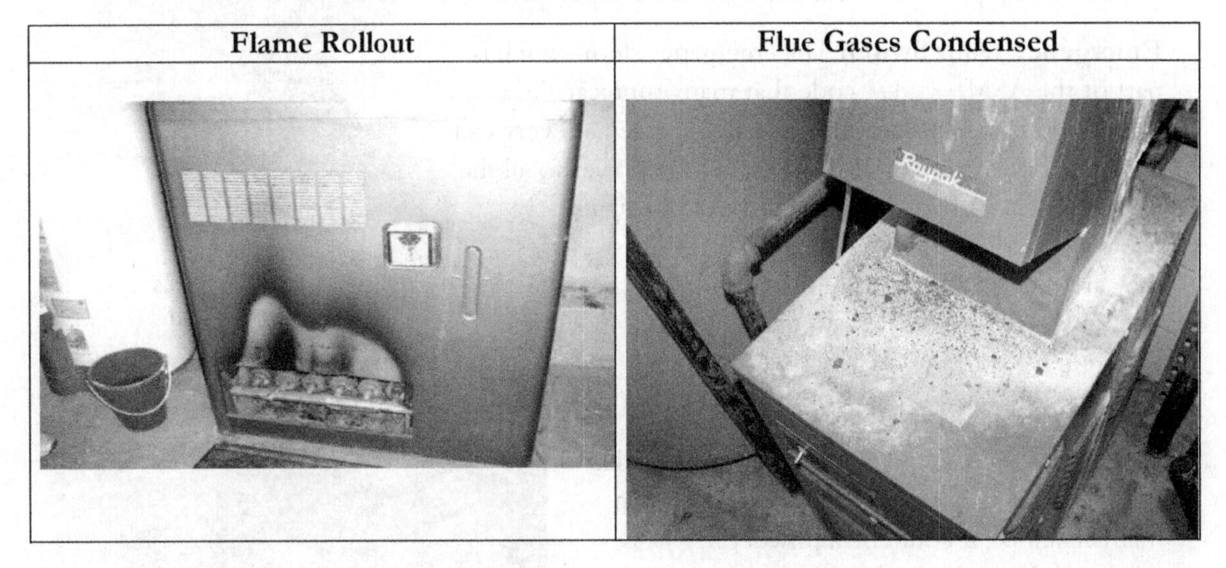	

Flame rollout? Flame rollout, as indicated by the discoloration on the boiler jacket, shows that flames and hot flue gases are not travelling the path that the equipment manufacturer wanted. If the flame is rolling out, it is a safe bet that carbon monoxide or other toxic gases are being produced. The cause of the flame rollout should be investigated. It may be from a negative condition in the boiler room, blocked flue passages, insufficient boiler water temperature, over firing the burner, or inadequate draft.

Negative conditions in boiler room? If the boiler room goes into a negative condition, this causes the flue gases to not draft properly. It does not take much, only – 3/100" W.C., to pull the flame from the burner. This is called back drafting and can produce toxic gases. Look for rust atop the boiler or water heater under the draft diverters. The rust is usually an indicator of negative conditions. I like to try simulating the worst case scenario to test for back drafting. If it is a house, I will…

- Open the hot water valve on a sink to start water heater.
- Start the clothes dryer.
- Start all exhaust fans including bathroom and kitchen exhaust fans.
- Close all windows and doors.
- Start oven if vented outside.
- Start boiler(s).

I verify the proper draft on the boiler and water heater using a draft gauge and test for carbon monoxide with my CO detector. It is a good idea to test for carbon monoxide around the boiler, water heater, and oven outlet.

If it is in a commercial boiler room, I will…

- Open several of the hot water valves to get the water heater started.
- Start the clothes dryer, if applicable.
- Verify all building exhaust systems are working.
- Start exhaust fan in boiler room, if any.
- Start air compressors that may be in boiler room.
- Verify combustion air dampers open.
- Start boiler(s)

I verify the proper draft on the boiler and water heater using a draft gauge and test for carbon monoxide with my CO detector. It is a good idea to test for carbon monoxide around the water heaters as well as the boilers.

Smell? When combustion appliances are not venting or firing properly, there will be a sour smell from the flue gases in the boiler room or basement. It is difficult to describe but will feel like it burns your nose and eyes. You also want to investigate any odors that smell like natural gas or propane in the room.

Boiler Room Temperature Extremely hot or cool boiler room temperatures could indicate a malfunction. See if you can identify the cause of the excessive temperatures. In one boiler room, I was greeted with a wall of heat as I entered. I found the cause of the heat was that the flames would roll out the rear of the boiler on an atmospheric burner. They had two boilers so I was able to shut the defective one off until repairs could be made. On another project, the automated combustion air dampers were frozen open causing the boiler room temperature to drop very low.

Air Conditioner If the boiler room has an air conditioner, the International Mechanical code requires that either the combustion air has to be directly vented to each fuel burning appliance or the room has to have a refrigerant monitoring system that will detect and alarm if it senses leaking refrigerant. Many of the older boiler rooms were installed before the current code was written and are grandfathered in.

Dissimilar Metals When you mix black iron pipe with copper, a reaction occurs called electrolysis or galvanic metal corrosion. This corrosion generates a small electrical current as a result of the two dissimilar metals and an electrolyte, water. This electrolysis can cause erosion of the metal, typically the black iron fittings. If you must join the two metals, a dielectric fitting should be used. The dielectric fitting separates the two metals and eliminates the chance of the galvanic corrosion.

CO Detector Is the room safe? Carbon Monoxide detectors should be used to verify that CO is not present in the room. You could use either the ones with an external sensing tube or the personal ones that clip on a belt. I would not use the plug-in ones for a service call as the reaction time is slow. A drawback to the plug in CO detector is the life expectancy of those are about five years. Some facilities also use combustible gas detectors.

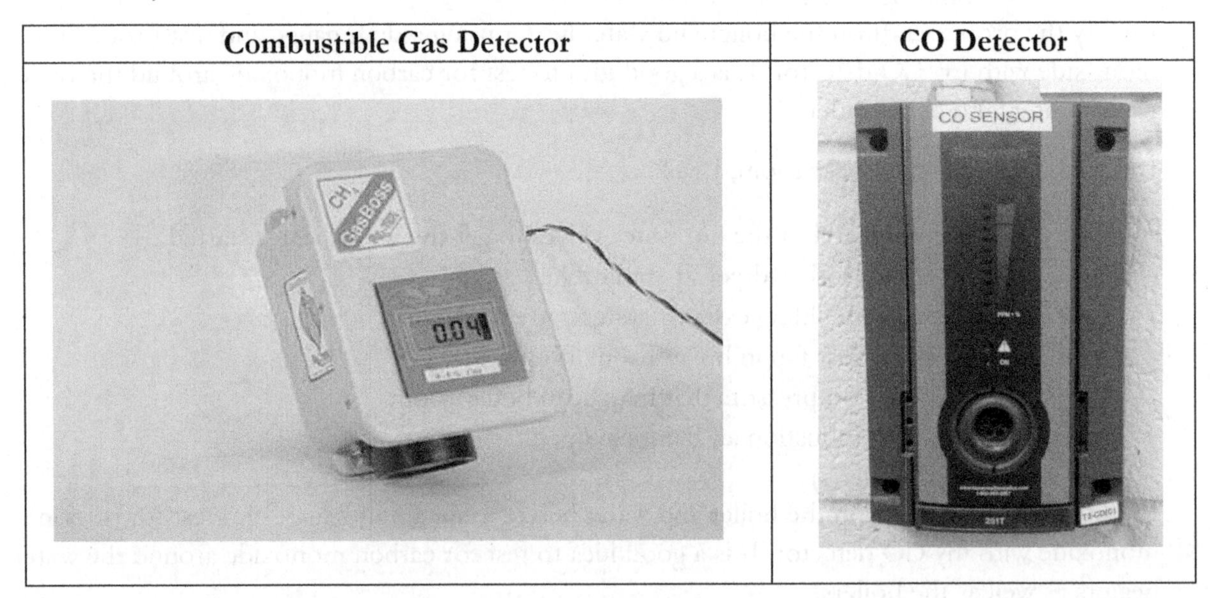

Listen for strange noises When entering the boiler room, listen for strange noises that will warn of potential dangers. Some of the noises I listen for would be:

Water dripping or leaking.

Electrical arcing may mean a short or bare wire.

Hissing like steam or gas escaping.

Loud ticking may be excessive scale buildup or a low water condition in the boiler.

Banging may indicate a boiler malfunction, especially a metallic banging.

Gurgling could mean air in the hydronic piping.

Popcorn or Rice Krispies Excessive scale or low water condition is sometimes thought to sound like either of these two items.

Excessive vibration could indicate a pump or motor malfunction.

Loud humming could mean a bound motor or pump.

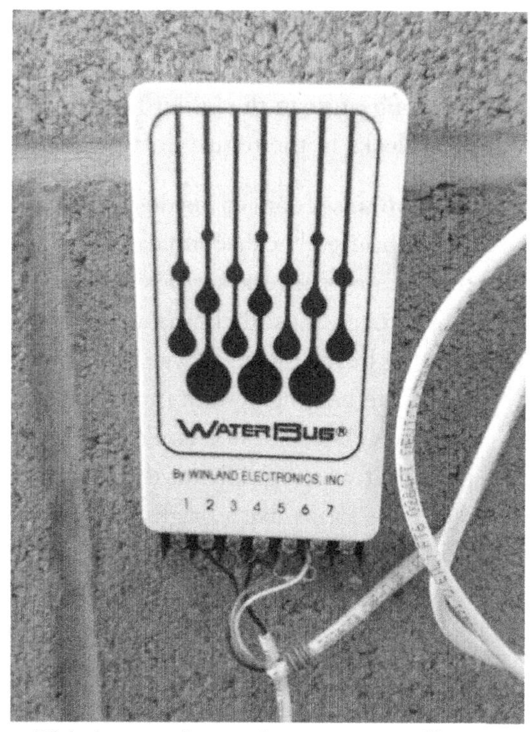

Water on boiler room floor may indicate a leaking pipe or boiler. Try to detect the origin of the leak. Water beneath a boiler could be from either a leak or flue gas condensation. A leaking relief could indicate that the pressure is excessive, the relief valve is getting weak, the piping to the expansion tank is plugged, or the expansion tank is flooded. *Did you know that some old timers used horse manure to plug leaking boilers?*

Fuzzy Logic While attending a meeting, an associate told me about a medical center where he serviced the boiler. The facility had a small maintenance staff that only worked daylight during the week. The staff was on call at night or on weekends. During a particularly cold weekend, a pipe froze and flooded the boiler room. It went unnoticed until early Monday morning when the maintenance staff returned. To eliminate that happening again, the maintenance staff wanted to install an alarm that would sound if water was present. This innovative maintenance staff found an old rocker switch and fixed it to the floor. Under the rocker, the staff placed Alka Seltzer tablets. If water was present, the tablets would fuzz and melt allowing the switch to drop and make the alarm switch. When asked about it, they said that since they were a hospital, they had plenty of Alka Seltzer.

Exhaust Fan In most instances, exhaust fans in a boiler room are your enemy. During the asbestos abatement project, all the insulation was removed from the steam piping. The boiler room was so warm that the glue holding the asbestos floor tiles in the classroom above the boiler room melted and the tiles curled up, creating a dangerous condition. In an effort to cool the boiler room, someone installed a thermostatically operated exhaust fan. The original combustion air louvers were sized for the boiler only and too small when the exhaust fan operated. The exhaust fan pulled the room into a negative condition and actually pulled the flue gases from the boiler. This filled the boiler room with a dangerous mixture of toxic gases. It also caused the boiler to soot, which exacerbated the issue. The technician's personal carbon monoxide alarm sounded as he entered the room. He was able to push the "Stop Boiler" button at the entrance and shut down the boilers. The steam piping had to be reinsulated and exhaust fan disabled.

If an exhaust fan is installed on a project, it should be electrically connected with a makeup air fan to assure that the boiler room cannot enter into a negative condition. Be careful of air handling units in the boiler room also as they can sometimes cause the room to be negative due to air leaks in the return duct.

Asbestos is a danger inside a boiler room as well as old houses because inhalation of asbestos fibers causes several serious illnesses including lung cancer, mesothelioma and asbestosis. If the existing heating system was installed prior to 1980, the heating system most likely contains asbestos. The asbestos could be on the flue or piping. It was typically used at the pipe fittings. Asbestos looks like a bright white flaky substance. Most commercial facilities had the insulation tested and should be able to inform you if any asbestos is in the boiler room. This should be known if you are working on the old boiler. When doing residential work, the homeowner may not know if asbestos is present.

Some of the older cast iron boilers used asbestos rope between the sections or on the flue collector to seal in the boiler flue gases. If you have any doubt, always have it checked.

Can the Boiler Breath? Many homeowners will have finished basements and relegate the boiler and water heater into a small closet and expect that they will work without problems. If the boiler uses indoor air for combustion, the International Fuel Gas Code 304.5 requires 50 cubic feet of volume for each 1,000 Btuh input of all the appliances. For example, if a home had a 150,000 Btuh boiler and a 40,000 Btuh water heater, the space required would equal

150,000 + 40,000 = 190,000 divided by 50 equals
9,500 cubic feet inside the room.

You would need a room roughly 30 feet by 40 feet for the appliances. If you find yourself in a small room like this, you should explain to the customer that the system requires combustion air. We have used combustion air fans that bring in air from the outside for these type rooms.

The picture to the right is a combustion air fan made by Field Controls. It will introduce combustion air to the boiler room on small projects and is wired in series with the boiler and water heater operation.

Courtesy: Field Controls

Boiler Maintenance, All

ASME CSD1 Code Requirements The American Society of Mechanical Engineers or ASME has a code, CSD1, that requires boilers to be properly maintained. The code CSD1 is for commercial boilers that are between 400,000 and 12,500,000 Btuh. The following are some of the requirements of the code:

CM110 *A systematic and thorough maintenance program shall be established and performed. Any defects shall be brought to the attention of the boiler owner and shall be corrected immediately.*

CM130 *Tests shall be conducted on a regular basis, and the results shall be recorded in the boiler log or in the maintenance record or service invoice.*

Documentation I urge you to document your findings when performing a boiler service call by including the unique numbers of both the boiler and burner as they may be from two different manufacturers. In addition to the model, I will include either the serial number or the National Board number for the boiler. The burner should have a unique number called the UL number. While it may seem to be a bit excessive, I had a client that replaced a burner on a boiler with a used one from an old boiler. He then called to tell us that the burner stopped working and we should repair it for free. Our documentation saved us on this.

Check the boiler in same conditions that it normally operates in I had a tech work for me that was claustrophobic and hated tight spaces. When he went into a boiler room, he would open all the doors because it helped him to breath he told me. The problem is that the boiler did not normally operate with the doors open. On one particular project, there was plenty of combustion air and the boilers ran great when the doors were open. Under the normal operating conditions with the doors closed, there was insufficient combustion air and the flames rolled out into the boiler room. Try to simulate the exact conditions that the boiler sees. I once had a snooty guy in a seminar that asked if he should turn out the lights when servicing the boiler because the boiler usually operates in the dark. Some people are so literal.

Going Dark. A client had some electrical work performed and the electrician wired the boiler into the light switch. It would only work when the boiler room lights were on. I lost a few hairs on that service call.

How Often Should We Check the Boiler? I prefer two visits per heating season but consult the installation and maintenance manual of the manufacturer and the owner's insurance company for their recommendations. I like doing a pre-season and mid-season check of the boilers. In the pre-season check, I will check all the safety controls and test the combustion flue gases to see if they are close to the proper fuel to air ratio settings. In the mid-season test, I test all the safety controls and perform the combustion analysis and adjustment. This is because the boiler and burner will operate longer to allow proper setup and adjustment. To properly adjust the fuel to air ratio of a burner, the burner should operate for at least 15 minutes to allow the flame to stabilize before any adjustments are made. If this is done in the fall at the preseason test, the boiler most likely will shut off on temperature and may take several hours to complete the test.

I look at boiler maintenance as a partnership with my clients and encourage my clients to accompany my service technician while doing the boiler service. In this way, they get to learn about their equipment as well as understand how the boiler operates. They could be your ally at the facility and save you from having to go out on a cold Friday night. I show the client the safety controls and the reset buttons and explain the proper settings on the boiler. I never had to worry about the client doing their own work after showing them how the boiler works. In contrast, it has led to more business for us with the client. While we cannot be there every day, the client should take some of the responsibility for their equipment. This includes the daily or weekly tasks that are suggested by the manufacturer or insurance company. There is nothing more embarrassing to the customer than when you find a switch off or a limit tripped and the repair takes five minutes. One of the things the customer will always ask is if we are going to charge for the service call or if we will be telling their boss. Rather than embarrassing the person, I suggest that we look over the boiler while we are there to make sure that everything is working. On another project, we found a high temperature limit control that was tripped. The client thanked us and wanted us to leave. I suggested that we check the boiler as the limit tripped for a reason. We found the operating control, which was set lower than the limit control, was defective which caused the control to trip. The operating control had to be replaced.

Checklist My mind tends to wander while doing tedious tasks so I like using a checklist when performing boiler and burner maintenance. The checklist assures that I check all the controls and components for the boiler. **See next page.**

Boiler maintenance logs are a great idea for any boiler room. The log contains a list of items that should be inspected on a regular basis on the boiler. These include safety devices on the boiler such as low water cutoff, operating and limit controls and relief valves. They are available on line or through the building owner's insurance company. It is like a checklist that is kept at the jobsite. I like to view the log when doing my boiler maintenance to see if any issues have occurred since my last visit. I encourage the client to write any issues on the log that they have seen on the boilers. If your state follows ASME CSD1, documenting that maintenance is performed is required by the code.

Heating Medium Most boilers are usually only one of two heating mediums, steam and hydronic. Some facilities, like industrial laundries use hot oil boilers to get elevated temperatures without a change of state. Steam is created when the water inside the boiler is heated to cause a change of state to a vapor. Hydronic systems will simply use warm water to heat the building. It is distributed to the building using circulators or pumps.

Types of Boilers There are two basic types of boilers, fire tube and water tube. A fire tube has the hot flue gases inside the tubes. The fire tube is surrounded by water. A water tube has the water inside the tubes and the hot flue gases surround the tubes.

Boiler Room Checklist	
All Boilers	**Hydronic Boilers**
Fuel Shutoff, Manual	Low Water Cutoff Primary
Fuel Shutoff, Electric	Water Feeder
Pilot Assembly	Flow Switch
Igniter	Circulating Pumps
Gas Pressure Regulator	Expansion Tank
Relief Valve	PTA Gauges
Flue	Piping
Induced Draft Fan	Water leaks
Wiring Connections	
Bearings Linkages	
Gas Train Venting	
Flame Safeguard	
Modulating Motor	
Modulating Control	
Low Fire Start Switch	
Combustion Air	
Operating Controls	**Low Pressure Steam**
Limit Controls	Low Water Cutoff Primary
Burner Air Flow Switch	Pump Control
Gas Pressure Switches	Low Water Cutoff Auxiliary
Refractory	Pressure Gauge
Main Burner	Siphon or Pigtails
Burner Mounting Plate	Boiler Feed or Condensate Tank
Blower Wheel	Boiler Feed / Condensate Pump
Blower Motor	Feed Water Valves
Boiler Base	Near Boiler Piping
Boiler Draft	Steam Leaks

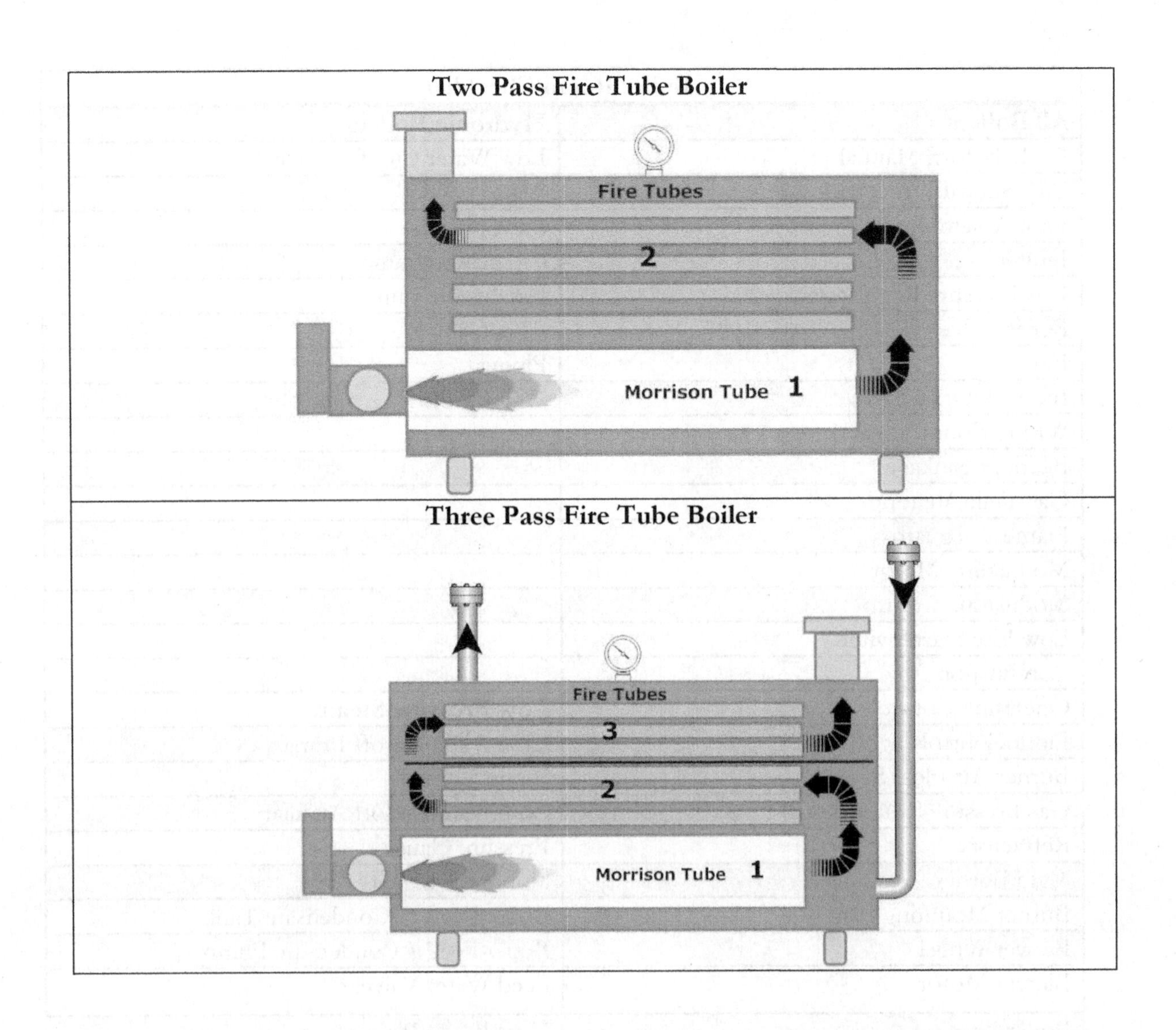
Two Pass Fire Tube Boiler
Fire Tubes
2
Morrison Tube 1
Three Pass Fire Tube Boiler
Fire Tubes
3
2
Morrison Tube 1

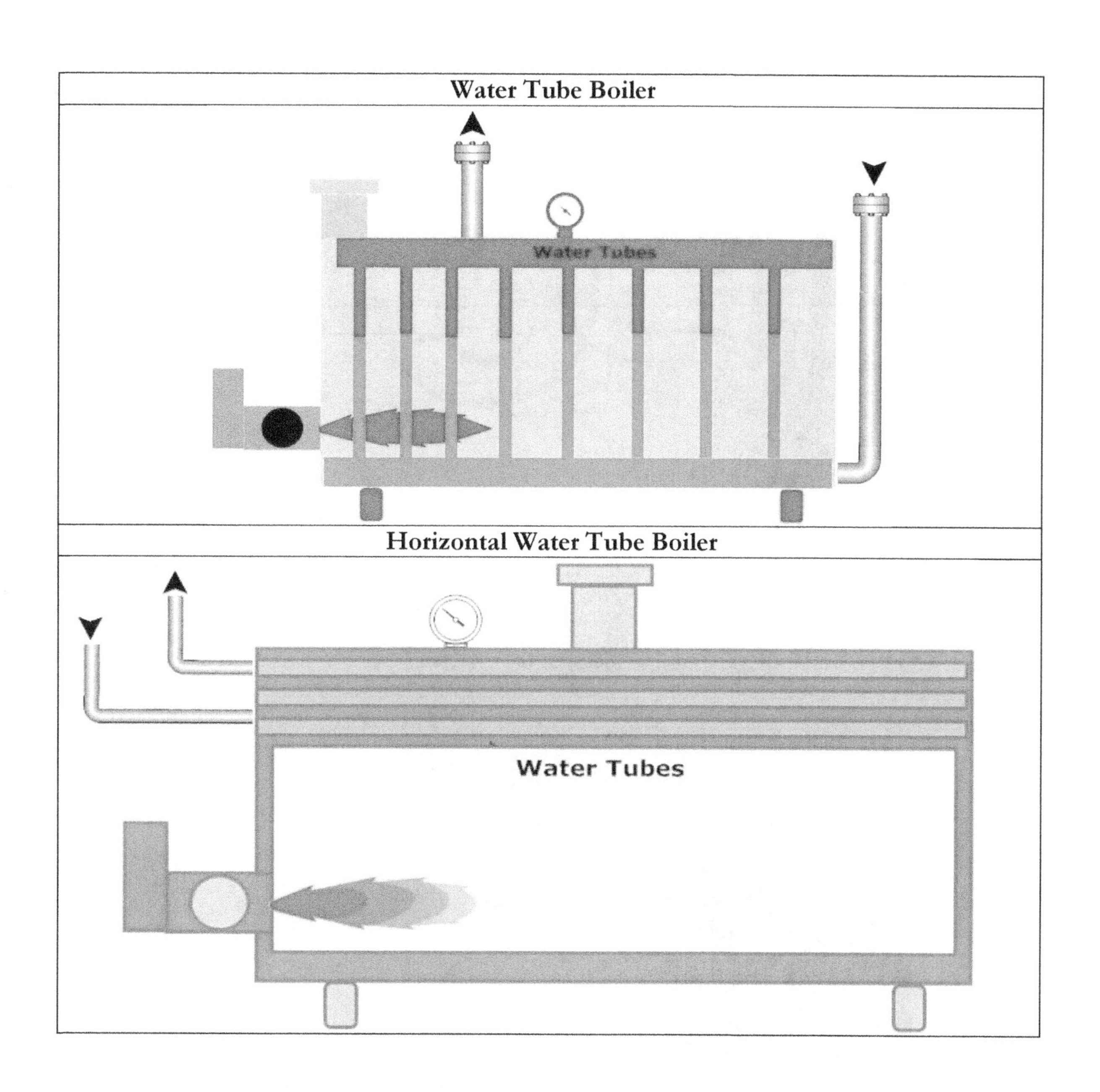
Water Tube Boiler
Water Tubes
Horizontal Water Tube Boiler
Water Tubes

Boiler Types

Steel Fire Tube

Cast Iron Sectional

Old Cast Iron Sections

Copper Water Tube Boilers

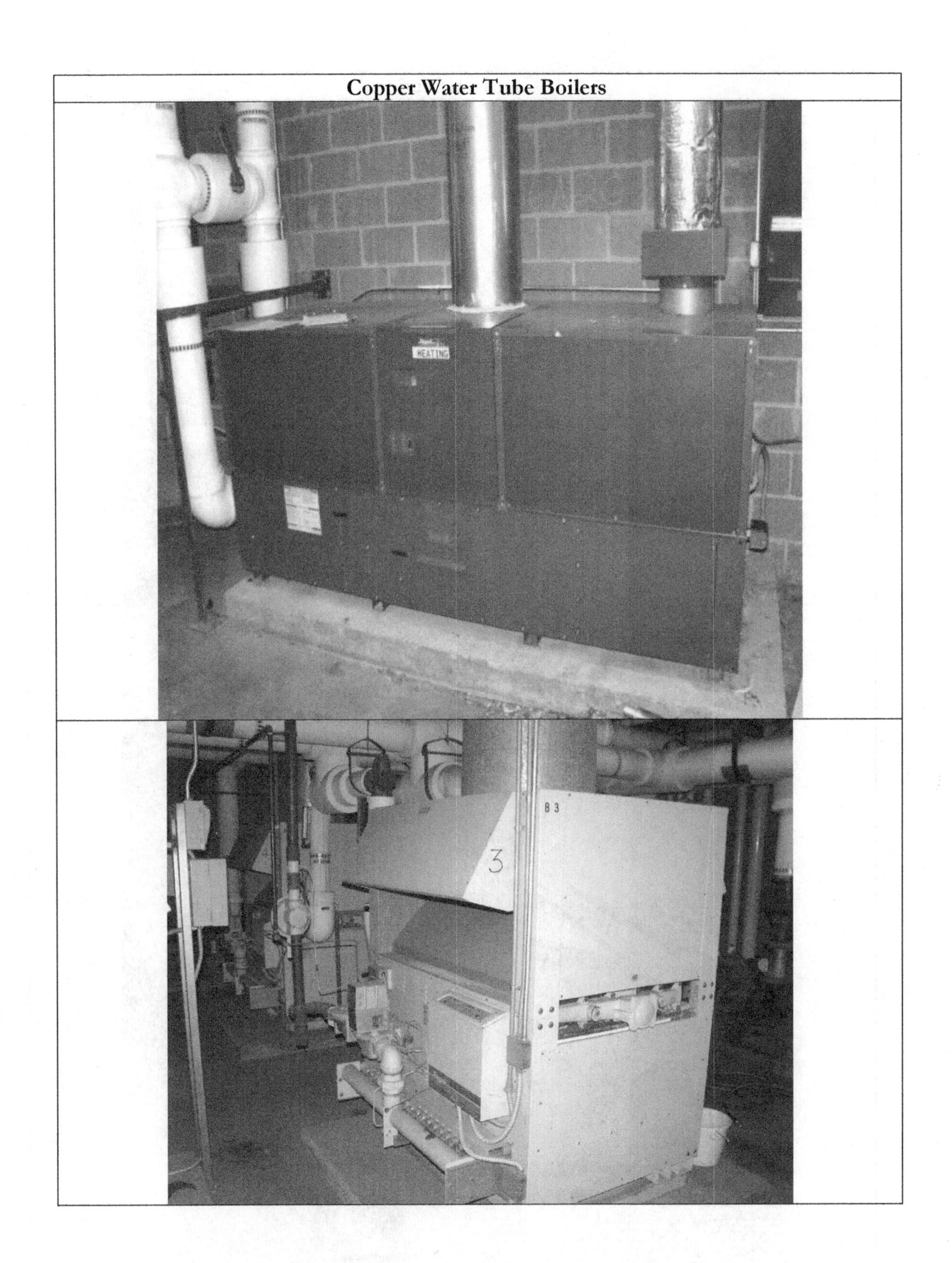

Steel Water Tube Boiler Cutaway Courtesy of Rite Boiler

Water Tube Boiler

Modular Boilers

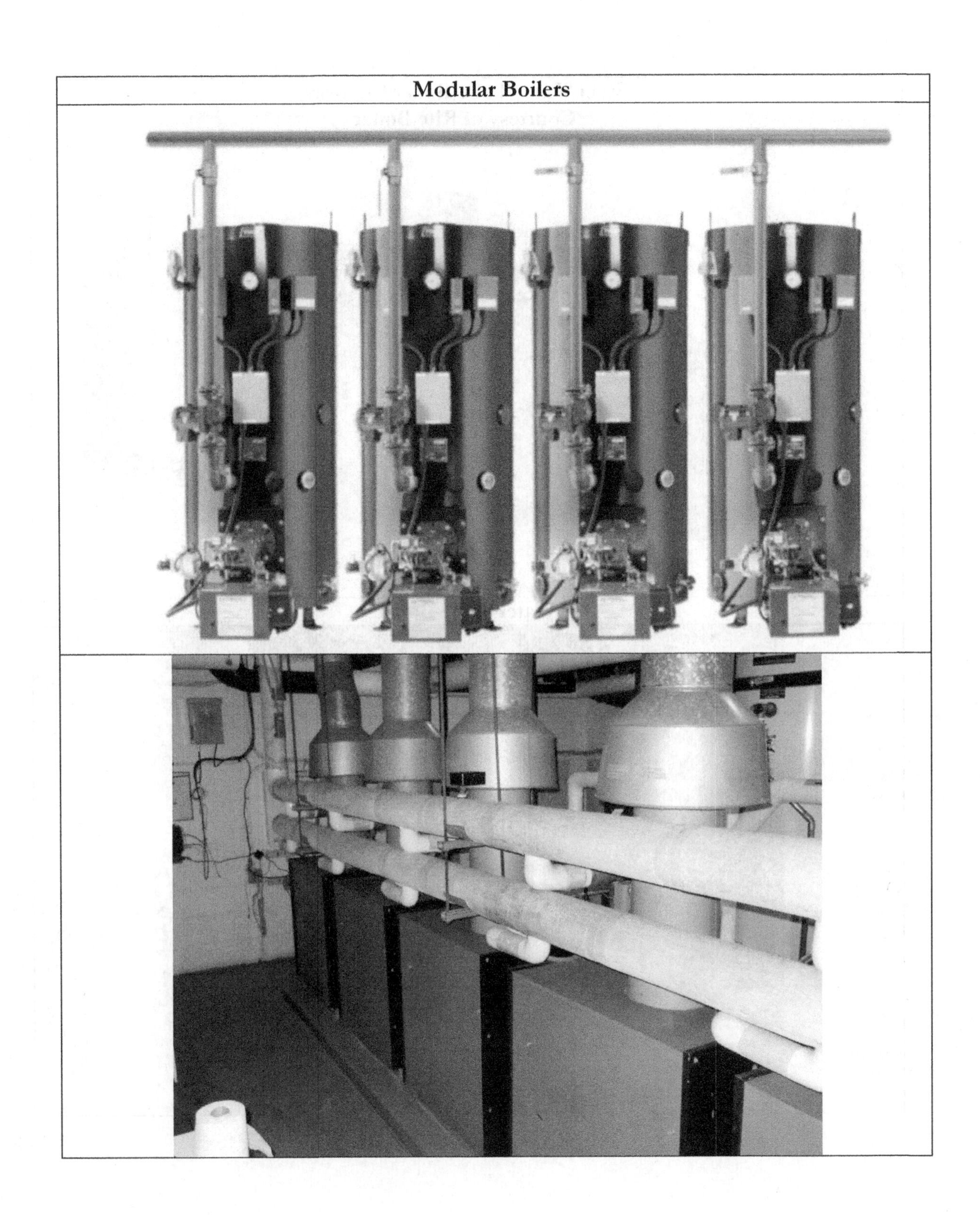

Hybrid System
Standard efficiency boiler on Left Condensing boiler on Right

Wall Mounted Boiler

Gas Burner Firing Types

Burners come in two styles; atmospheric and power. The atmospheric burner has a quicker response because it does not have the pre and post purge on the firing cycle like power burners do. Boilers with atmospheric burners tend to be less efficient than the ones with power burners. The atmospheric burner uses gas pressure and chimney draft to mix the fuel and air.

A power burner uses an internal fan to introduce air to the combustion process. While it requires more electricity to burn the fuel, the added electrical costs are offset by the lower fuel costs.

Power Burner

Most of the newer boilers are multi position burners. The following are the types of burner firing options you will find in the field.

On-Off The burner starts when there is a call for heat. Once the setpoint of the operating control is reached, the burner will shut off. An On Off burner is usually the least expensive type of burner but it allows wider swings of the temperature or pressure inside the boiler.

Low High Off (LHO) This burner starts at low fire for smoother ignition and once the flame is established will travel to high fire and stay there until the call for heat has ended. This is a two position burner.

Low High Low (LHL) This burner starts at low fire and when the flame is established will travel to high fire. It will stay at high fire until it reaches the setpoint on the firing rate control. At that point, the burner drops to low fire until the call for heat ends or the temperature or pressure at the sensing point of the firing rate control drops low enough to drive it to high fire again. The firing rate control is usually set at a lower setpoint than the operating and limit control. For example, the firing rate control on a typical hydronic boiler will be set at 175 degrees, the operating control set at 180 degrees F and the manual reset limit set at 200 degrees F. This is a two position burner. According to Honeywell, a Low High Low burner will be 15% more efficient than a modulating burner.

Modulating This burner will start at low fire and then travel to high fire until it reaches the firing rate control setpoint. At that setpoint, it will vary the firing rate anywhere from high fire to low fire depending upon the signal from the firing rate control. This is a multi-position burner.

What is Turndown? If you are like me, that is what happened when you asked your children to clean their rooms. When it pertains to burners, it is quite different. Burner turndown is the percentage of the high fire rate that the burner can safely reduce to. The lowest firing point is called Low Fire. Let us assume that we are going to purchase a 100,000 Btuh boiler and the supplier has the three following options:

Boiler #1 has a 2-1 turndown

Boiler #2 has a 5-1 turndown

Boiler #3 has a 10-1 turndown.

What does that mean?

Boiler #1 - 2-1 Turndown means that the boiler can reduce firing from 100,000 to 50,000 Btuh.

Boiler #2 - 5-1 Turndown means the burner can reduce firing rate from 100,000 to 20,000 Btuh.

Boiler #3 - 10-1 Turndown means the burner can reduce firing rate from 100,000 to 10,000 Btuh.

The higher the turndown, the less the boiler cycles. This helps when you need a consistent temperature or pressure in the boiler. The firing rate of the burner is controlled by either the firing rate or modulating control.

Comparative Fuel Values

To get 1,000,000 Btu's you need the following

Fuel Source	1,000,000 Btu's
Natural Gas @ 1000 Btu/ cu ft	1,000 Cu ft
Coal @ 12,000 Btu/ lb	83.333 Lb
Propane @ 91,600 Btu/ gal	10.917 Gal
Gasoline @ 125,000 Btus/gal	8.000 Gal
Fuel Oil #2 @ 140,000 Btus/gal	7.194 Gal
Fuel Oil #6 @ 150,000 Btus/gal	6.666 Gal
Electricity @ 3,412 Btu/kWh	293.083 Kwh

Combustion Air

Combustion Air Openings The journeyman technician wanted to teach me, his apprentice, the importance of combustion air in boiler rooms. We measured the openings and the technician reached into his tool bag for a hammer and walked to the glass window and broke the glass out of one of the panes of a window. He smiled and said, "There you go." The owner saw this and was livid. The technician justified breaking of the window by saying, "You didn't have enough combustion air." After we were escorted from the jobsite and the irate customer called the boss, it was explained to me in between curse words that this was not the proper way to increase the combustion air for the room but you can decide if this direct method works for you.

Why do we need combustion air? Each burner requires a certain amount of air for proper combustion, typically 12-15 cubic feet of air for every cubic foot of gas burned. There are up to three types of supply air to a boiler; primary, secondary, and tertiary or dilution air. The primary air is used to support the flame. The secondary is what is used to add turbulence to allow proper burning. The tertiary or dilution air is used to make sure the flame and heat do not leave the boiler prematurely. It allows the room air to enter the flue when the stack is warm and draft is high. Power burners may not have the dilution air. If the combustion air is inadequate, the burner will attempt to find air somewhere, anywhere. A local post office contracted us to check their boiler. We found the combustion air louver was covered with cardboard because it was too cold in the room and the smokers congregated there. I noticed that the flue pipe for the water heater was riddled with holes and white powder. Since the combustion air louver was blocked, the boiler would get its combustion air by pulling it from the water heater flue. In addition to pulling combustion air, it also pulled flue gases from the water heater flue into the boiler room.

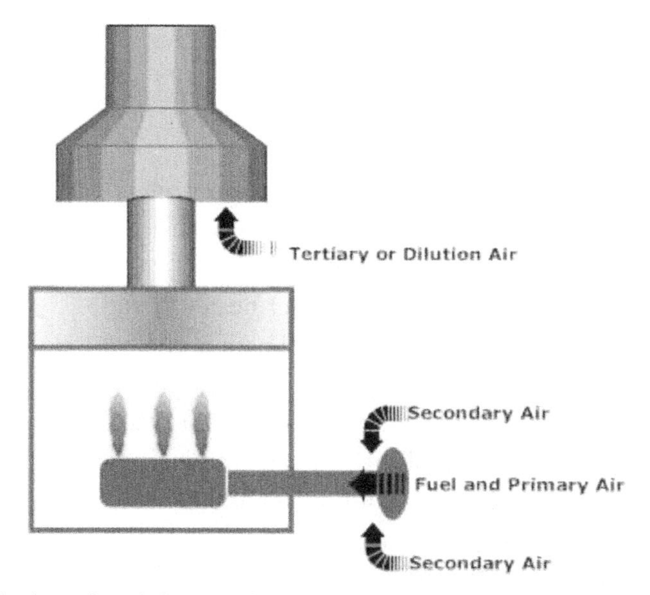

Verify the size of the existing combustion air for the equipment room. When you measure the area of the combustion air louver, confirm with the louver manufacturer what the "A_K" factor is for the grill. The A_K factor or Area Factor is the actual free area of the grill. A rule of thumb is that metal grills have about 75% free area and wood grills have about 25% free area. For example, a 10" x 10" louver equals 100 square inches. A metal louver would have 75" free area and a wooden louver would have 25" free area. When calculating the sizing requirements, add all the fuel fired equipment in the room. This would include the water heater, pool heater, and boilers.

Combustion Air Openings The following are some guidelines for help in sizing the combustion air louvers.

Number of openings required = 2

- Each boiler room should have two openings. One should be within one foot of ceiling and the other opening within one foot of the floor.

The reasons for two openings is twofold. The first is that two openings are less likely to both be blocked by leaves, grass clips, or garbage. The second is that the high and low openings provide natural ventilation inside the boiler room.

Combustion Air Sizing	
Combustion Air Introduction Method	**Size Required**
Direct Openings	1" Free area for each 4,000 Btuh
Horizontal Openings	1" Free area for each 2,000 Btuh
Vertical Openings	1" Free area for each 4,000 Btuh
Mechanical Ventilation	1 cfm per 2,400 Btuh

Using Indoor Air for Combustion Boilers seem to be shoe-horned into the smallest spaces possible and it could affect the operation of the boiler or water heater. If the boiler uses indoor air for combustion, the International Fuel Gas Code 304.5 requires 50 cubic feet of volume for each 1,000 Btuh input of all the appliances. For example, if a home had a 150,000 Btuh boiler and a 40,000 Btuh water heater, the space required would equal

150,000 + 40,000 = 190,000 divided by 50 equals 9,500 cubic feet inside the room.

You would need a room roughly 30 feet by 40 feet for the appliances. If you find yourself in a room like this, you need to explain to the customer that the system requires combustion air.

Louver Screen Size - The International Mechanical Code requires louver screen holes to be not less than 1/4" and not more than 1/2" in size. Check to make sure the outside grill is not covered or blocked. In many facilities, we see grass clippings and leaves blocking the openings.

Ah, the old water line in front of the combustion air louver ploy. I have seen many boiler rooms where the plumber has installed a water pipe in front of the combustion air louver. Be careful that the pipe does not freeze.

Are the combustion air openings too small?

Let us see if the combustion air louver will be adequate for our project. The building has two 10" x 10" openings with metal louvers. That equals 100 square inches each or 200 square inches total. According to our A_K estimates, we have 75% free area or 75 inches free area each or a total of 150 inches free area. The grill is a direct connection to the outside. Based upon 4,000 Btuh per one inch free area, our openings would be adequate for 600,000 Btuh total input. What if the heating system is larger than that? Well, you have a couple of options. One is to direct vent the combustion air for your boilers. This would involve installing combustion air duct directly from the outside to the burner. If you choose to do this, you may have to insulate the duct as it will sweat when the cold air is introduced into the warm boiler room. The second option is to either increase the opening or add another opening through the outside wall. A third option would be to install a makeup air fan. If you choose mechanical ventilation, you would need 1 cfm per 2,400 Btuh. If your new heating system is rated at 1,000,000 Btuh and your water heater is 40,000 Btuh, you would need fan capable of delivering 433 cfm. (1,000,000 + 40,000 = 1,040,000 divided by 2,400 CFM = 433 CFM) The makeup air fan should be interlocked with the burner so that the burner does not operate until the makeup airflow is verified. In most instances, you could use a differential pressure switch.

Using Combustion Air Fans I saw a project where the engineer designed the combustion air supply using a combustion air fan for each boiler. Each smaller fan would be electrically connected to the boiler it would serve and would only operate when its boiler started. It kept the boiler room warmer and reduced the electrical costs.

Boiler Room Ventilation In some instances, you may still need more ventilation air to the room even though you have adequate combustion air. Johnston Boiler recommends 2 cfm per boiler HP just for boiler room ventilation. If your new boilers have combustion air directly vented to burner, the room may not get adequate ventilation air and it could get stuffy with elevated levels of CO2 or carbon dioxide. In addition, the boiler room always seems to be the repository of cleaning chemicals and tools. Chemicals should not be stored inside the boiler rooms. The off gassing of the chemicals may mix with the flame and create dangerous conditions.

Automatic combustion air dampers Part of your inspection should be to verify the operation of the combustion air dampers. ASME CSD1 code stipulates that there should be an interlock on the driven damper that assures it is open before firing the boiler.

The boiler room is too cold I visited a boiler room on a no heat call and found that the boiler room was cold, really cold. As a matter of fact, you could see your breath in there and the temperature was so low that the gas valve would not open. I was afraid that pipes would freeze. I measured the combustion air louvers and they were correct for the boilers and water heaters in the room. The problem was that they were fixed position louvers and open all the time. To get the boiler running, we had to put portable heaters in the room. To resolve the problem, I suggested either automatic combustion air dampers that would only open when there was a call for heat or thermostatically controlled permanent heater for the room.

What if the owner blocks the combustion air? I have been on many jobsites where the owner has covered the combustion air louvers with cardboard or plywood because the boiler room felt cold. It is an unsafe condition that could lead to injuries or even death. So, what do you do? I will document it by taking a picture of the blockage and then explain to the customer the importance of combustion air, explain why it is so dangerous, and offer to remove the cardboard or wood. Most will comply once they realize that it could impact the safety of the people in the building. I will also make a note on the service report about what I did and take another picture when completed. On one project, the customer refused to let me remove it as the boiler room was where the smokers in the building congregated. I kindly asked the customer to sign my service report that I advised him that the situation was unsafe. I also shut off the boilers and told him that he could do whatever he wanted to do after I walked out of the building. We sent a follow-up letter to the customer detailing the dangers so we would be not held responsible. The customer found another boiler service company and that was fine with me.

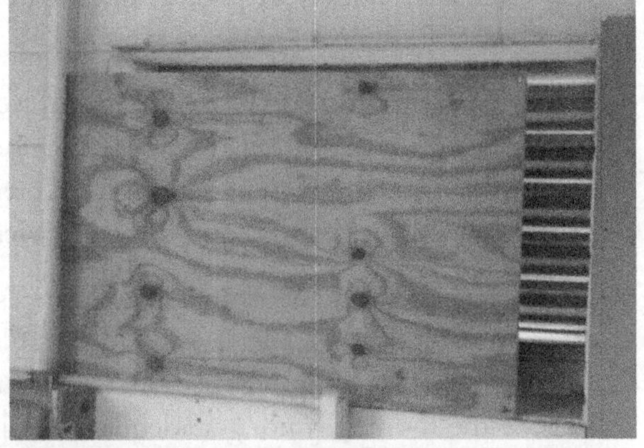

Smoking in the boiler room. For many years, the boiler room was the only place in the building where the smokers could light up. During a service call at a local high school, I caught a whiff of marijuana smoke in the boiler room. I looked into the adjacent tunnel and did not see anything. I told my contact, the Maintenance Director, for the school that I thought some students may be in the tunnel smoking the drug. He laughed and said, "Nope, It's the teachers. How else do you think they can deal with the little monsters?"

The following is a table showing the Btu capacity of direct connect metal combustion air louvers. It is based on 75% free area of the metal louvers and 4,000 Btuh per inch of free area. Add all combustion burning equipment in the room and measure the outdoor air louver.

Btuh Capacity of Direct Vent Metal Combustion Air Louvers								
	The numbers below are based in thousands. Please add ,000 to the number for total Btuh capacity.							
Inches	**Width in Inches**							
Height	**12**	**28**	**36**	**48**	**60**	**72**	**84**	**96**
12	432	864	1,296	1,728	2,160	2,592	3,024	3,456
18	648	1,296	1,944	2,592	3,240	3,888	4,536	5,184
24	864	1,728	2,592	3,456	4,320	5,184	6,048	6,912
30	1,080	2,160	3,240	4,320	5,400	6,480	7,560	8,640
36	1,296	2,592	3,888	5,184	6,480	7,776	9,072	10,368
42	1,512	3,024	4,536	6,048	7,560	9,072	10,584	12,096
48	1,728	3,456	5,184	6,912	8,640	10,368	12,096	13,824
54	1,944	3,888	5,832	7,776	9,720	11,664	13,608	15,552
60	2,160	4,320	6,480	8,460	10,800	12,960	15,120	17,280
Based on direct connect metal louvers with 75% free area and 4,000 Btuh per inch of free space.								

This chart shows the CFM required for mechanical boiler room ventilation using fans. CFM Required @ Various Boiler Btuh			
Btuh	CFM	**Btuh**	CFM
50,000	21	**1,100,000**	458
75,000	31	**1,500,000**	625
100,000	42	**2,000,000**	833
200,000	83	**2,500,000**	1,042
300,000	125	**3,000,000**	1,250
500,000	208	**3,500,000**	1,458
700,000	292	**4,000,000**	1,667
900,000	375	**4,500,000**	1,875

Direct venting of combustion air

Some boilers require combustion air vented directly from the outside. This will be covered in a later chapter but some things to be cautious of when using direct vented combustion air for the boilers are:

Condensation When the cool air is brought in from the outside to the warm boiler room, the air will condense inside the duct. This could cause the combustion air duct to sweat. In addition to a housekeeping issue, the moisture will cause the duct to rust and corrode. If using duct for the combustion air, it should be insulated.

Dirt Since the air is coming from the outside, it may contain dirt and anything else that is blowing around outside the boiler room. A project we sold had massive construction going on outside the boiler room. This dust and dirt from the parking lot ended up inside the burner. We had to clean the burner several times during the construction. What about filters? You may ask. Well, the filters could also plug from the dirt and starve the burner for the much needed combustion air.

Critters Make sure the intakes are covered with something to keep the critters out. It may be something as simple as a screen. We have found dead birds and squirrels inside the combustion air ducts.

Intake Location The intake location is important as it could pull in the flue gases that are being exhausted out the same wall.

Natural Gas Piping

- Gas fittings should be malleable iron and not black iron.
- Many gas companies do not permit bushings to be installed in gas lines. They prefer bell reducers.
- Threaded fittings greater than 4" shall not be used except where approved. IFGC 403.10.4
- Piping shall not be installed in or through a ducted supply, return or exhaust, or a clothes chute, chimney or gas vent, dumbwaiter or elevator shaft. IFGC 404.1
- Piping in concealed locations shall not have unions, tubing fittings, right and left couplings, bushings, compression couplings and swing joints made by a combination of fittings IFGC 404.3

IFGC = International Fuel Gas Code

Combustible Gas Leaks It is a good idea to test for gas leaks on the gas train components. While I prefer using an electronic leak detector, I have used the tried and true soapy water test for the components. To use the soapy water test, mix dishwashing liquid in a container with water. Apply the soapy water to the pipe fittings with a small paint brush and if a large bubble or several smaller ones appear, there is a gas leak. Look for leaks on the pipe threads and around each component. A service technician that worked for me used his lighter to check for gas leaks. He did not work for me very long. Please do not use a lighter to check for gas leaks as it is dangerous and unprofessional.

Combustible Gas Detector

The combustible gas detector we use makes ticking noises as it senses flammable gases. The higher the concentration, the quicker the clicks, reminding me of a Geiger counter used to test for radioactivity. My service manager, a practical joker, would start the detector in crowded elevators and turn up the sensitivity to start the loud clicks. People would always spin around and ask what was wrong and he would tell them it was nothing to worry about, which would make them worry. In case you were wondering, the detectors can sense the gas in flatulence, another of his practical jokes.

Gas Piping The gas supply to the boiler should have a "Dirt Leg" or sediment trap to catch dirt and other items that might be in the gas line. The dirt leg should be 3" minimum in length. The pipe to the burner should be at a 90 degree angle to the incoming gas. This is to protect the gas train components. Some contractors will install the dirt leg in the run of the pipe. This does not work as well as the one on the left because the contaminants will be entrained in the gas flow.

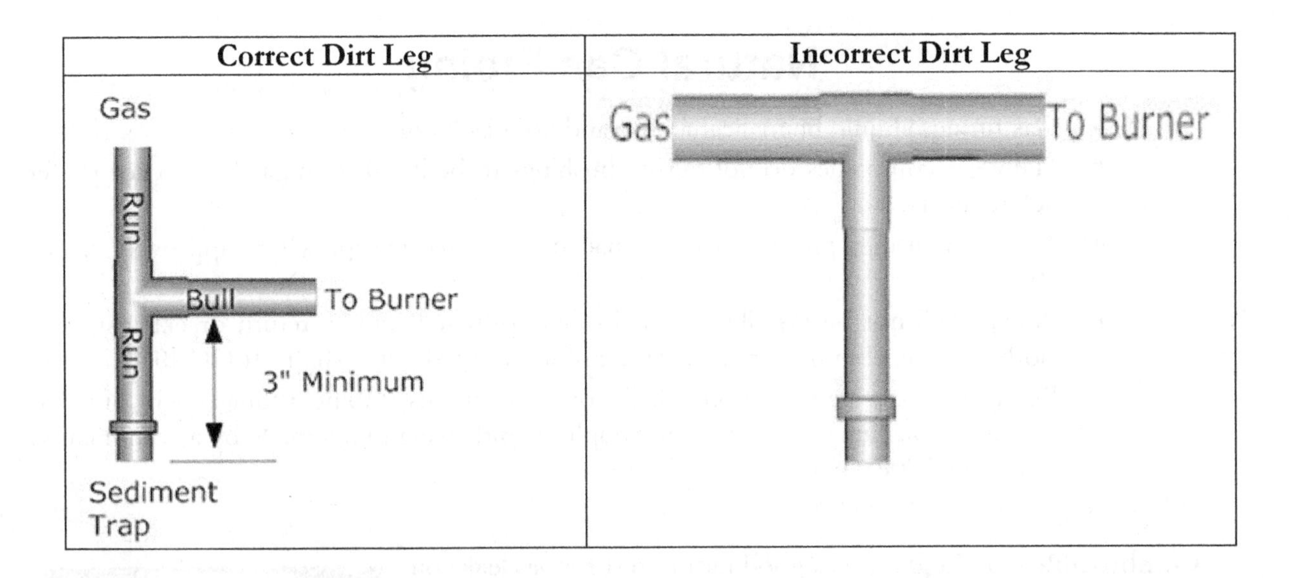

Gas Pressure It is always a good idea to keep the upstream manual gas valve closed until the gas pressure is verified to be in the proper range. I was asked to perform a startup of several boilers in Atlanta, GA and when I arrived, the upstream gas valve was open and the supply gas pressure was 5 pounds and the boiler gas train was designed for less than one pound. All the gas pressure regulators locked up and I could not get the boilers working. The contractor had to install a regulator to drop the pressure down to the ½ pound setting and replace all the boiler regulators as they were ruined.

Gas Quality When performing boiler service, you are at the whims of the company supplying the gas to the building. We service the boilers at my church and the gas line had a leak in the pipe as there was moisture and dirt in the gas line and it caused flame failures and other maintenance headaches. We installed an extra dirt leg and strainer to capture some of the dirt and moisture in the supply pipe. When we spoke to the gas supplier, they denied the dirt and moisture was in their piping.

Teflon Tape on Gas Piping Some manufacturers of boilers, burners and gas valves will void their warranty if Teflon tape or pipe sealant is used on the installation. Please review the installation instructions of the new equipment to see if it mentions the use of Teflon as a thread sealant. If it does, there are other types of pipe thread sealants that can be used that do not contain Teflon.

The following is from the Webster Engineering installation manual:

"Warranties are nullified and liability rests solely with installer when evidence of Teflon is found." I may sound like Mr. Obvious but that seems clear cut that they do not want Teflon used on the gas train. Many fuel oil burners do not want Teflon used on their components as well.

Gas Train

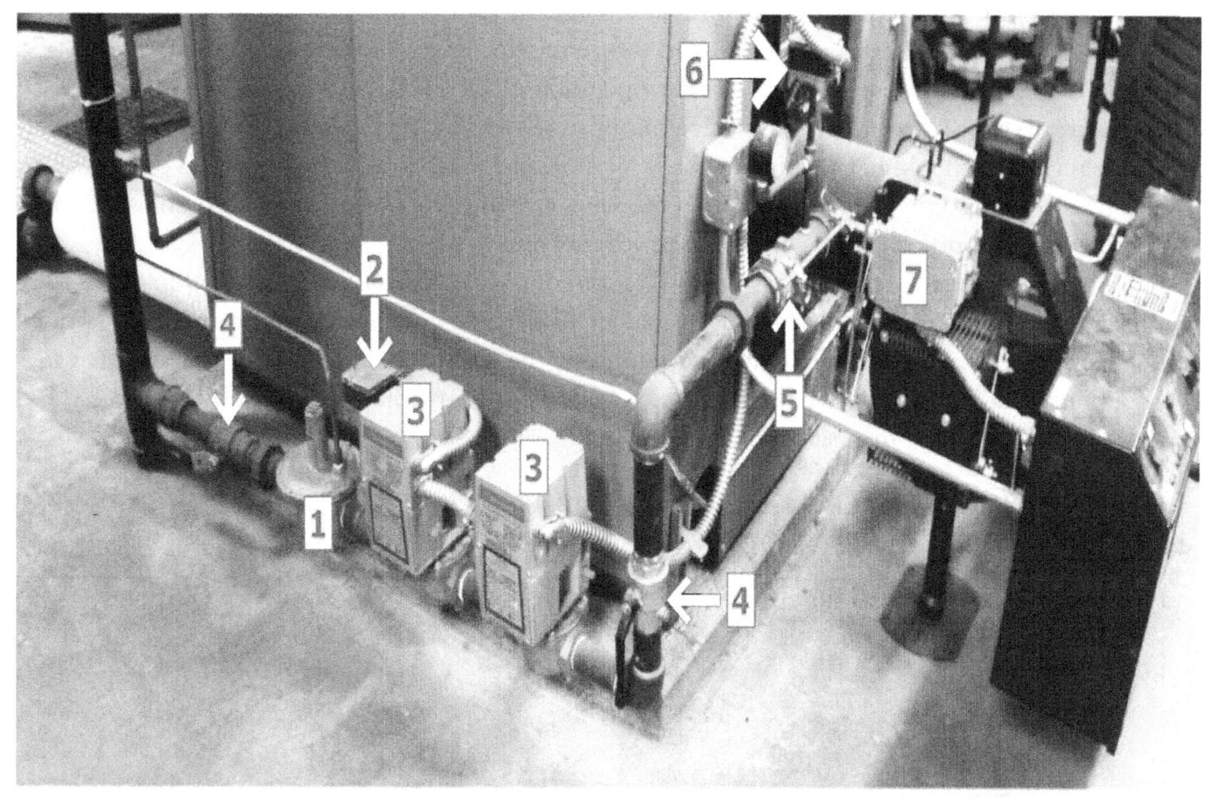

The gas train is comprised of several components in the fuel piping for the boiler that assures safe operation of the boiler or burner. The main components in the gas train include the following:

1 Gas Pressure Regulator

2 Low Gas Pressure Switch

3 Safety Shutoff Valve(s)

4 Manual Shutoff Valve(s)

5 Firing Rate Valve

6 High Gas Pressure Switch

7 Modulating Motor

Gas Train

1 Gas Pressure Regulator The gas pressure regulator assures that the burner receives the proper amount of natural gas to safely and efficiently fire. The location of the gas pressure regulator is important as it could affect the operation of the burner. It should be installed downstream of the first manual gas valve but upstream of the first electric gas valve. The outlet of the regulator should have a nipple that is 10 times the diameter of the pipe to avoid turbulence and noisy combustion. The gas supply to the regulator should not be interrupted unless servicing or isolating the burner. When servicing the gas pressure regulator, a gas pressure gauge or manometer that reads in Inches of Water Column or "W. C." should be used. If you use a pressure gauge that reads only pounds, it may be difficult to get an accurate reading. One pound of pressure equals 28" w.c. To adjust the setting on the gas pressure regulator, remove the cap covering the adjustment screw. To increase gas pressure, turn adjustment screw clockwise. To decrease gas pressure, turn screw counter-clockwise.

Important: If you adjust the gas pressure regulator, you should test the fuel to air ratio of the burner using a calibrated combustion analyzer.

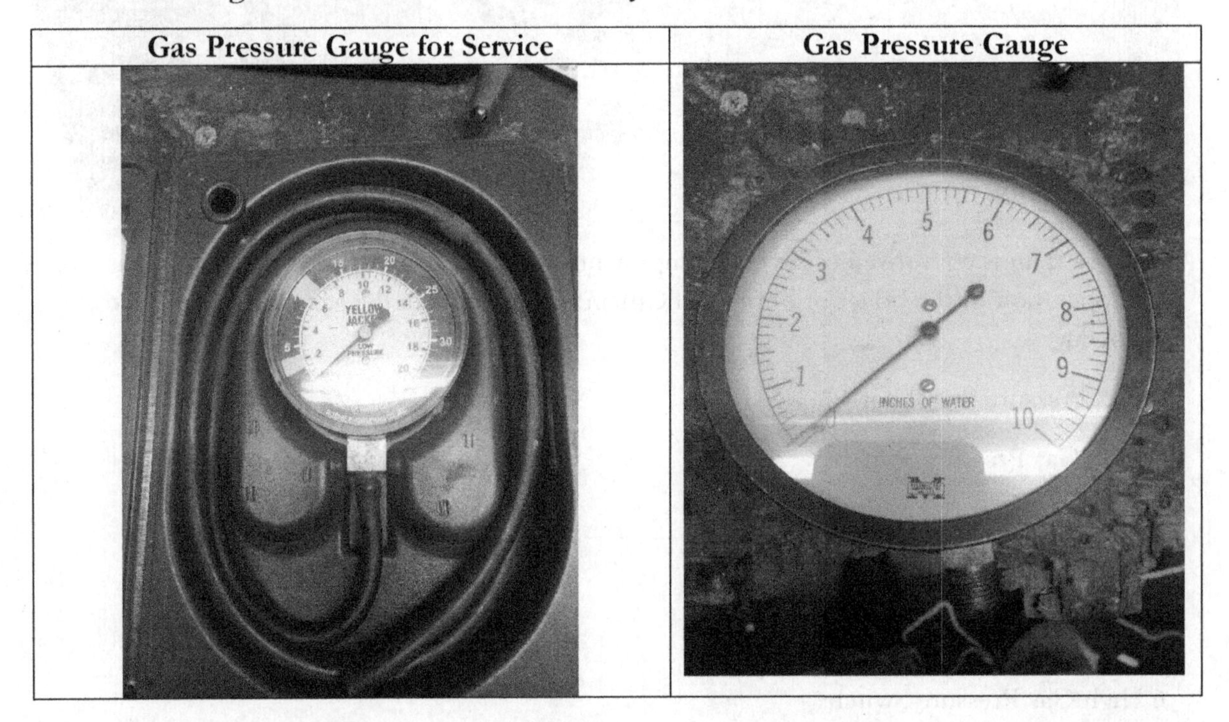

Gas Pressure Gauge for Service	Gas Pressure Gauge

Correct Gas Pressure? The gas pressure works hand in hand with the boiler draft. The boiler draft can change the firing rate of the burner without you realizing. I was on a jobsite and the gas pressure was spot on to what the manufacturer recommended. When I looked at the flame, it was being pulled up into the boiler. I tested the draft and found it was ten times the amount that the manufacturer wanted. Instead of -0.05" w. c. , it was -0.5" w.c. Once I adjusted the draft, the burner was grossly over-firing and I had to lower the gas pressure.

Regulator Service As regulators age, they may lose some of the accuracy. On many models, there is a spring under the adjustment screw that can be replaced. The springs are available in several pressure ranges. Some manufacturers will use a color code to show the pressure ratings of their springs. The following is the pressure rating for the Maxitrol RV series regulators that are one of the most common for burners.

Spring Color	Inches W. C.	kPa	Spring Color	Inches W. C.	kPa
Plated	3-6"	0.75-1.5	**Violet**	4-12"	1-3
Orange	4-8"	1-2	**Green**	5-15"	1.25-3.7
Blue	5-12"	1.25-3	**Red**	10-22"	2.5-5.5
Brown	1-3.5"	0.25-0.9	**Yellow**	15-30"	3.7-7.5
Plated	2-5"	0.5-1.25	**Black**	20-42"	5-10.5
Pink	3-8"	0.75-2			

Many of the gas pressure regulators are rated for an incoming pressure less than ½ pound or 14" w.c. If the pressures exceed the rated inlet gas pressure, it will lock up the regulator and could destroy it.

Blocked Vent Every year, we get "No Heat" calls due to plugged vent pipes. A plugged vent will sometimes mask itself as a defective gas valve or gas pressure regulator as you will have gas pressure and power to the valve but no gas pressure downstream of the valve on gas valves with internal regulators.

To test if the vent is plugged, disconnect the tubing from the regulator and see if it opens. If it does, the vent piping may be plugged. If you have a plugged vent, you could temporarily operate the burner by disconnecting the vent tubing. I would only leave it disconnected until I can clear the blockage and never walk away with it that way.

No Cutting in Line We had a project where the gas regulator was downstream of the first electric valve. When the electric valve was de-energized, the gas pressure regulator went wide open. On the next call for heat and the electric valve opened, the gas pressure regulator was wide open. It over-fired the boiler and flames shot out the side of the boiler until the regulator adjusted the pressure. To resolve the problem, we had to relocate the regulator upstream of the electric valve. On smaller boilers, you would typically see a range of 4-7" w.c. (w.c. = water column) Larger commercial boilers will usually see a range of 7 to 14" w.c. Industrial boilers may be rated for several pounds of gas pressure. On smaller, residential boilers, the regulator is part of the gas valve. The regulator may require venting to the outside. When venting the gas train components outdoors, I would suggest installing an insect guard on the discharge. This may be something as simple as a screen. Insects will sometimes block the outside termination of the vents. The vents cannot be combined in the same pipe with the normally open vent valve on a block, block and bleed gas train.

Gas Pressure Switches

Low and High Press Gas Cut-off

2/6Gas Pressure Switch Some burners have gas pressure switches to assure that the gas pressure is within the proper parameters. There are two basic gas pressure switches, low and high. The low gas pressure switch is located on the gas train downstream of the gas pressure regulator. If the gas pressure is below the setpoint, it will trip and shut off power to the burner. The setting will usually be set to trip when the gas pressure is 50% of the burner manifold gas pressure.

The high gas pressure switch is located downstream of the last electric safety shutoff valve and is commonly mounted on top of the burner head. It will interrupt power to the burner if the gas pressure is above the setpoint. It is usually set to trip if the gas pressure reaches 150% of the burner manifold gas pressure.

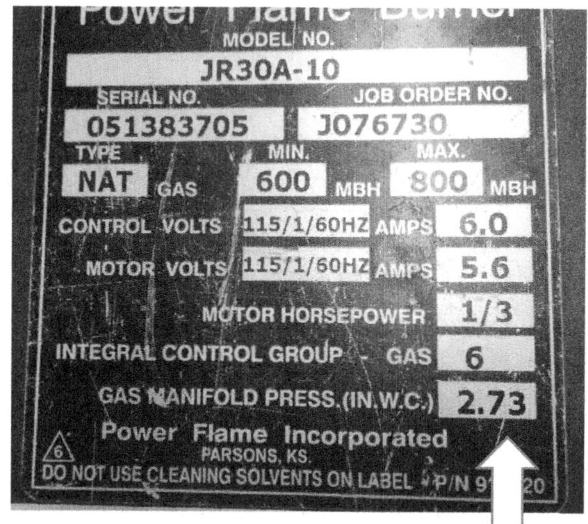

Both of these controls will most likely have a manual reset button that will require resetting if they are tripped. Normally, I prefer testing a safety control in its normal operating conditions but in the case of a gas pressure switch, I will do it differently. The low gas pressure switch should open if the upstream main gas valve is closed. The testing of the high gas switch is a bit trickier. I do not like adjusting the gas pressure regulator to increase the gas pressure to the switch as it affects the fuel to air ratio. To test these switches, I will adjust the setpoint down on the control and note where the control trips. I will then adjust the switch back to the normal position. If the burner manufacturer called for a gas manifold pressure of 3", the low gas pressure switch would be set for 1.5" w.c. and the high gas pressure switch would be set for 4.5" w.c.

Manifold Pressure	Low Gas Switch Setting	High Gas Switch Setting
3	1.5" W.C.	4.5" W.C.
4	2" W.C.	6" W.C.
5	2.5" W.C.	7.5" W.C.
6	3" W.C.	9" W.C.
7	3.5" W.C.	10.5" W.C.
8	4" W.C.	12" W.C.
9	4.5" W.C.	13.5" W.C.
10	5" W.C.	15" W.C.

3 Safety Shutoff Valve(s) These gas valves, also called Electric Shutoff Valves, are electrically operated and normally closed. They require electrical power to open. In a typical heating season, these electric valves may be cycled thousands of times. You will most likely find that the valves use 24 or 115 VAC although some old burners used 220 VAC. Some older gas valves also used millivolts and are called Powerpile. More on those below. On large commercial boilers, you may have two gas valves. They come in all sorts of styles and sizes and range from a solenoid valve to a slow opening motorized valve. These valves should be inspected regularly and verified that they work properly. Check for gas leaks around the gas valve as well as if they allow gas to leak through the valve. Many of the gas valves have plugs in the bottom to allow you to install pressure gauges to test operation.

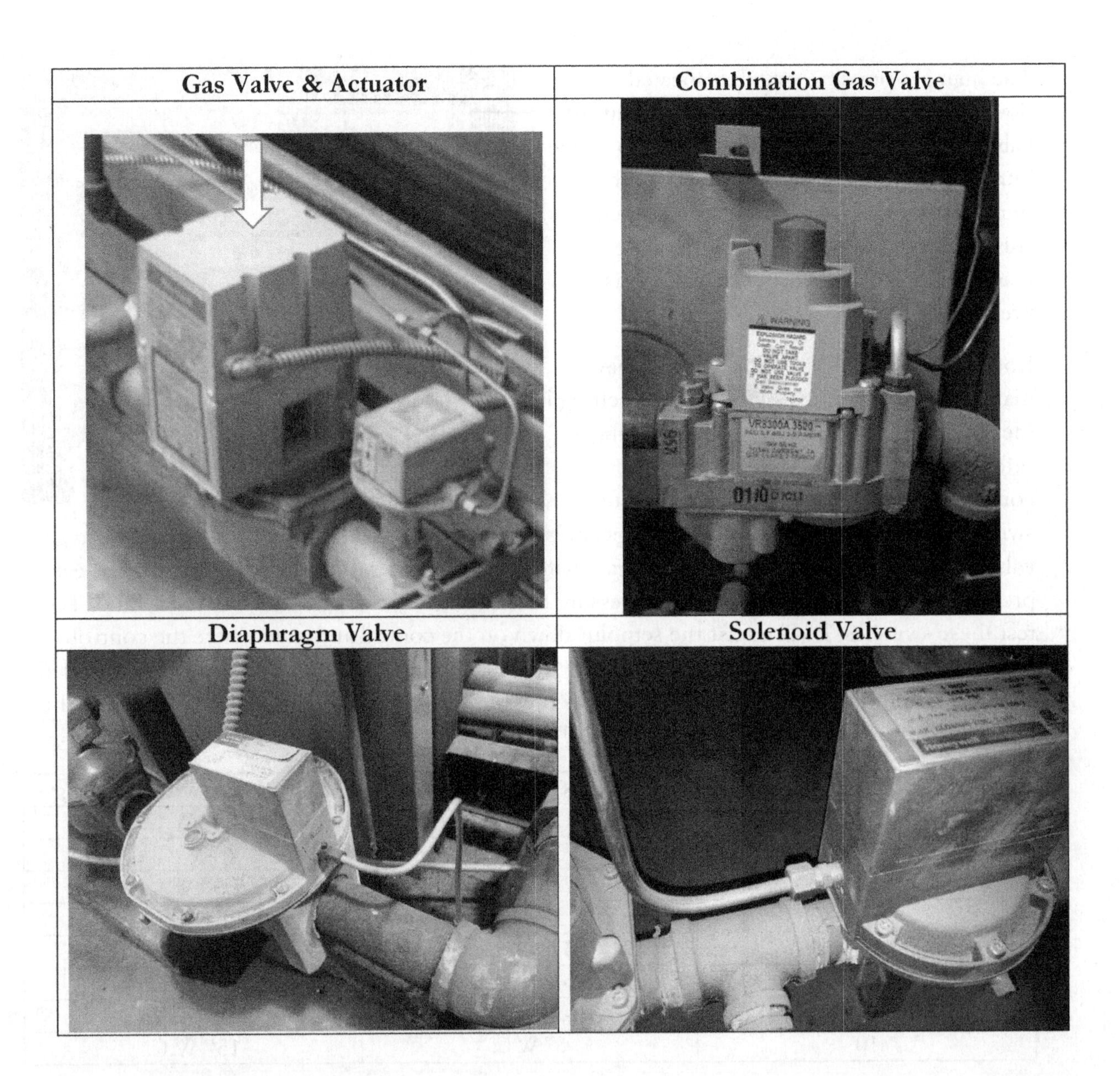

Check the following items:

- Wiring connections
- Gas Leaks
- Hydraulic Fluid Leaks (usually visible in valve window)

To test the operation of the electric gas valve, you will have to use a gas pressure gauge on both sides of the gas valve and verify that it opens when electricity is present. If both pressures are the same, the gas valve is open. If the inlet shows gas pressure and the outlet does not, the valve is not open. This assumes that we have gas pressure to the valve inlet.

Some diaphragm valves use an adjustable bleed fitting on the vent connection. This turns the normally On Off valve into a slow opening valve. This looks like an ordinary brass ¼" fitting that is used to vent the valve but it has an adjustment screw that controls the speed of the valve opening. If you are having problems with rumbling, this device may solve your problem. If the screw is closed, the valve will not open.

Honeywell Smart Valve The Honeywell Smart Valve, designated with an SV prefix is an all-inclusive gas valve that will provide gas flow control and main burner sequencing. They are sometimes confusing to troubleshoot due to the complexity. Honeywell distributors have a test harness (Part # 395466/U) that will simplify the diagnostics of the control and allows you to measure the flame signal.

Millivolt System I had a millivolt or Powerpile system on the one pipe steam boiler at my first house. I loved it because we had frequent winter power outages and the boiler worked without external power. It is ingenious how these systems work. The pilot flame heats a pilot generator or Thermopile that looks like a thick thermocouple, about ½" thick. The Thermopile will produce between 500 to 750 millivolts. Each millivolt is 1/1,000 of a volt. The typical Powerpile gas valve requires a minimum of 100 millivolts. The Thermopile generator should have a minimum of 325 millivolts. If replacing the thermostat for one of these systems, make sure that it can be used for this type of system. Some thermostats with heat anticipators will not work with a Powerpile system.

Bleed Valve On older gas trains, many were piped using a Block, Block, and Bleed configuration. In between the two safety shutoff valves was a tee. The bull of the tee was piped to a solenoid valve called the Bleed Valve. The outlet of this Bleed valve was piped outside. The purpose of this valve is to vent any gas that may leak past the first safety shutoff valve. This bleed valve is normally open. A union should be installed above the vent valve to allow inspection of the valve and to facilitate leak testing of the valve. The life expectancy of a solenoid valve is about ten years. They should be checked regularly to make sure they do not leak. A client called me to say that his fuel costs went up drastically one month and asked us to see if we could see a problem. His maintenance man had replaced the normally open vent valve with a normally closed one. Anytime that the

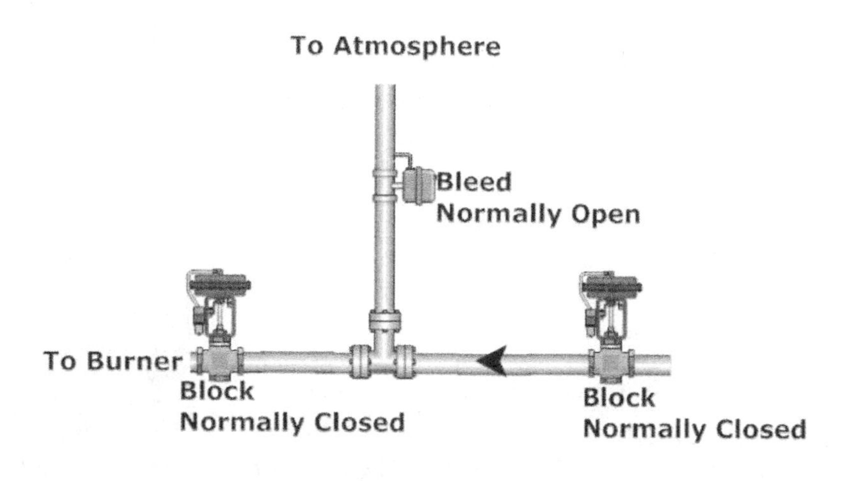

burner was on, the vent valve was open spilling natural gas from the 1" pipe. His gas bills dropped dramatically when it was replaced with the right type of valve.

Normally Open means that the valve is open without power and closes once power is applied to the valve. Normally Closed means that the valve is closed except when power is applied. Most heating valves are Normally Closed.

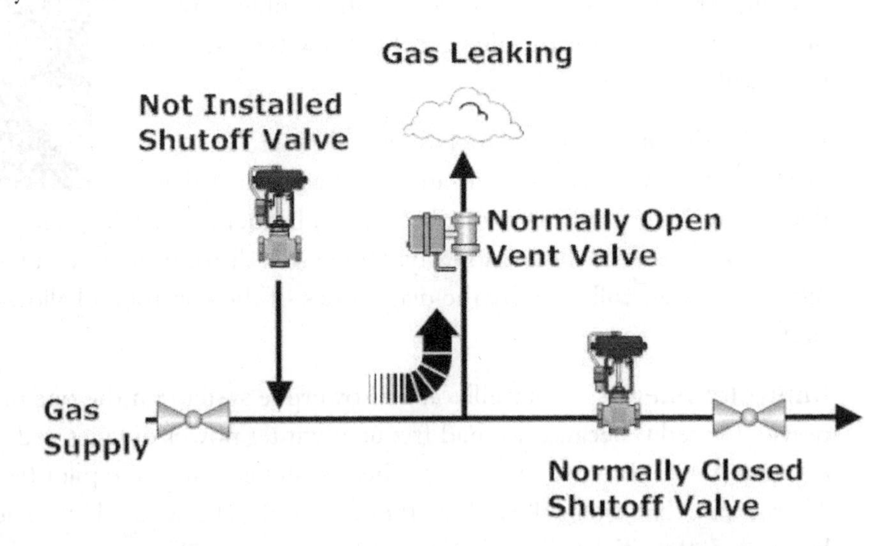

The sequence of operation for this type of gas train is as follows:

Upon a call for heat, the burner ignites the pilot. Once the pilot operation is verified, the two block valves open and the normally open vent (Bleed) valve closes. When the call for heat is satisfied, the block valves close and the bleed or vent valve opens.

I thought it was an extra We sold a boiler to a contractor that had never installed a commercial boiler. I asked if he would like to have me stop by to discuss the installation and he declined saying that he knew how to install a boiler. He called to inform us that the job was ready to be started. He was anxious to get paid. When we arrived at the project, I looked at the gas train and was shocked. The first electric gas valve was missing. Luckily, the manual gas valve was shut off or else gas would have poured from the normally open vent valve. I asked what happened to the gas valve and he answered, "I thought they sent a spare."

Fuel Shutoff Actuator On large commercial burners, the gas valve may require an actuator that is mounted on the gas valve. The actuator is what is used to open or close the gas valve. Inside some of these actuators, are switches that used to verify the proper position of the gas valve.

Dangerous Gas Valves Some of the older gas valves were very dangerous and could cause injury or even worse. The first valve is referred to as NaK valve. This valve contains Sodium Potassium Alloy which could ignite spontaneously if exposed to water or even humid air. Uh, guess what we find in most boiler rooms; water and humid air.

The second valve has a manual switch that allow you to open the valve without electricity. It is almost like having a manual valve. Both of these valves should be replaced with modern safer valves.

NaK Valve	Manual Opening Valve

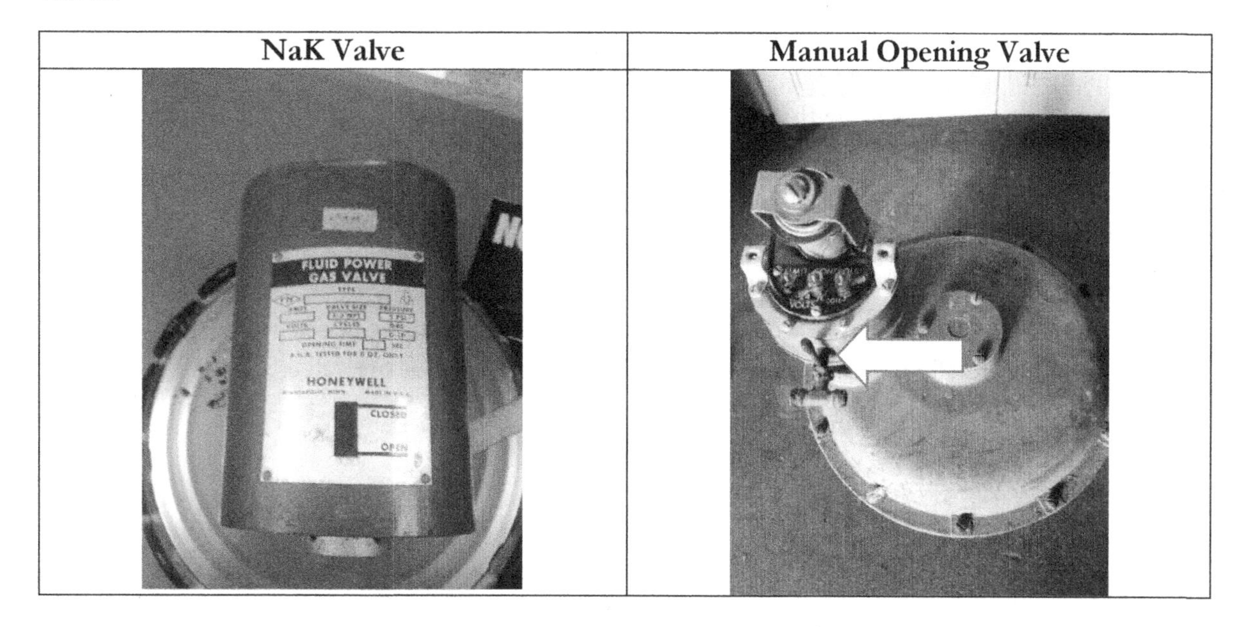

Did you know...

that old timers would use horse manure to seal leaking boilers?

Gas Valve Leak Testing

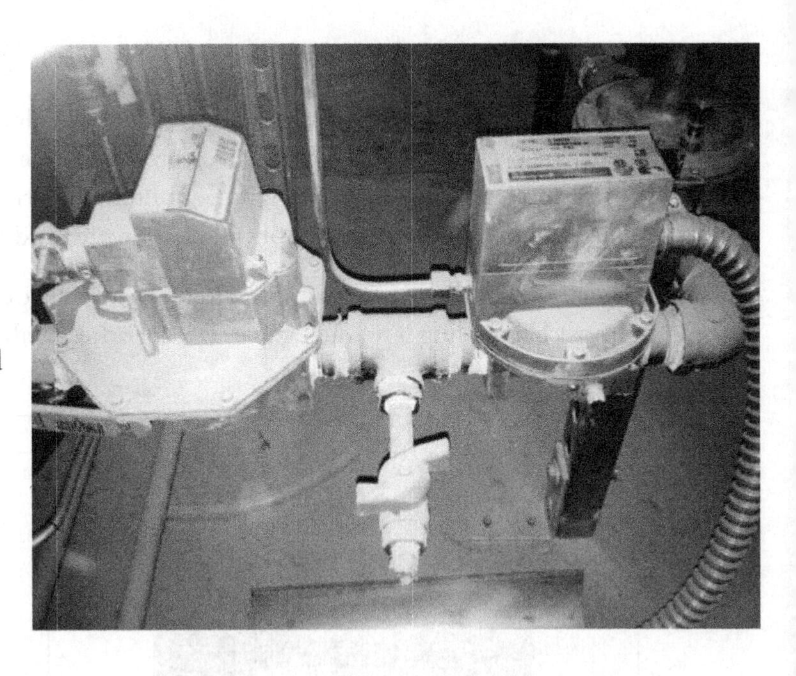

*ASME CSD1 - American Society of Mechanical Engineers Controls and Safety Devices for Automatically Fired Boilers Code requires that you should be able to leak test the gas valve for gas leakage through the valve. The picture to the right shows how one contractor piped the gas train to allow for leak testing.

It stuns me that there is even an acceptable amount of leakage through a gas valve. And how do we test for a leaking gas valve? Expensive detector with a digital readout? No, we count bubbles. Really!

According to ASME CSD1, as well as most gas valve manufacturers, provisions should be made in the gas train to allow for leak testing of the electric safety shut off valves (SSOV). This is

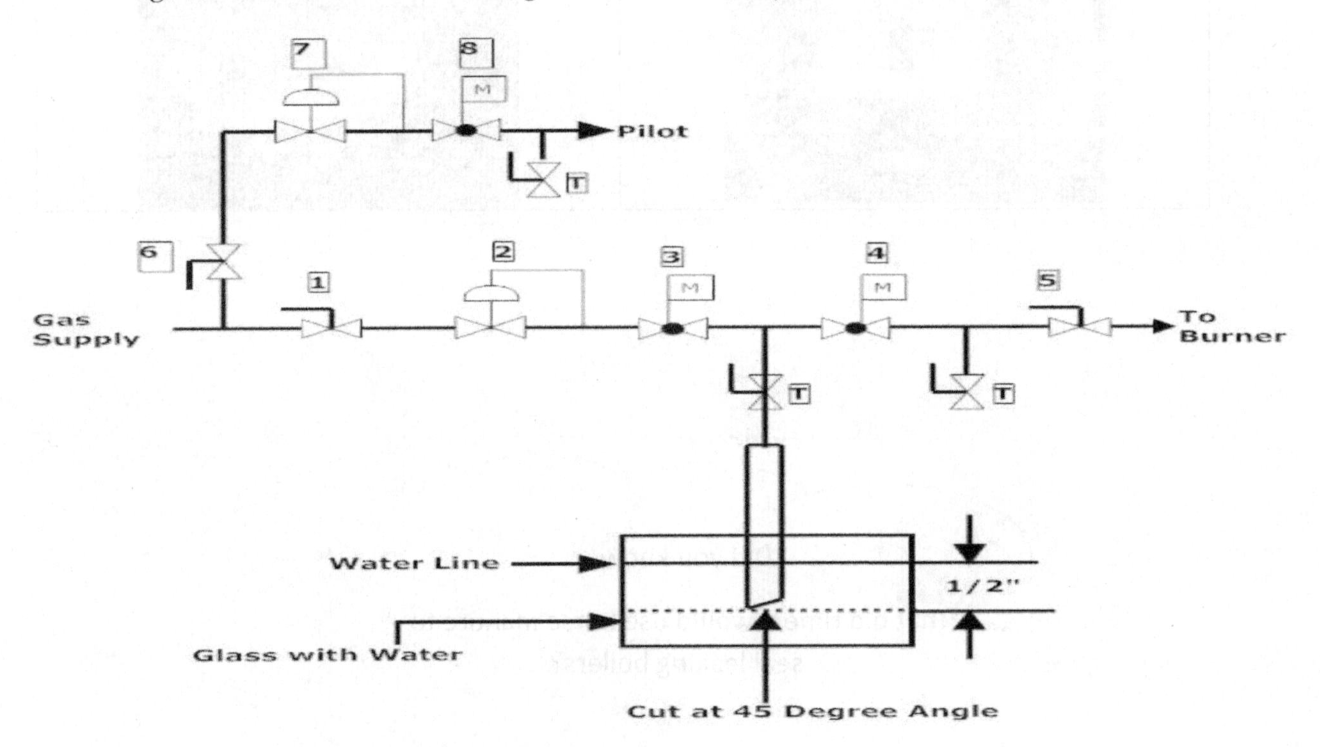

1	Upstream Manual Gas Valve
2	Gas Pressure Regulator
3	Safety Shutoff Valve
4	Safety Shutoff Valve or Blocking Valve
5	Downstream Manual Gas Valve
6	Manual Gas Valve- Pilot
7	Pilot Gas Pressure Regulator
8	Pilot Safety Shutoff Valve
T	Test Ports These may be part of valve

usually a ¼" tapping in the downstream nipple after the gas valve. CSD1 and the manufacturers recommend testing the valves at least once per year. If there are two electric SSOV's, then both have to be tested. The pilot solenoid valve should also be tested at least once per year. Some valves will have a downstream tapping as part of the valve. If the gas train contains a bleed valve, I would suggest testing this at the same time. This valve is piped to the outside and unless it is tested, a leak may never be found. Some service organizations use a manometer to leak test the gas valves. When testing the valves, refer to the manufacturer's recommendations. To meet the U.S. requirements, leakage must not exceed ANSI Z21.21, Section 2.4.2. It is based on air at standard conditions and limits leakage to a maximum of 235 cc/hr per inch of seal-off diameter. This is not the same as pipe diameter. The following is the maximum bubble count for the valve sizes.

Valve Size (Inches)	Allowable Leakage (cc/hr)	Maximum Bubbles per 10 second test.
¾"	458	16
1'	458	16
1 ¼"	458	16
1 ½"	458	16
2"	752	26
2 ½"	752	26
3"	752	26
4" & Larger	1,003	35

4 Fuel Shutoff, Manual In most gas trains, there are two manual gas valves. One is upstream of all components. This valve usually has a pipe plug and is typically where the pilot is attached. See picture to the right. Please note the gas flow when installing this valve. The pilot connection is upstream of the main gas valve. There is another manual valve that is downstream of all other components. This will allow you to isolate the gas train components. To test this valve, you simply close or exercise the valve. Verify that the valve does shut off and there are no leaks of gas from the valve into the boiler room. Be cautious when opening or closing these valves as they could be frozen in place if they have not been exercised in a while. This valve is crucial for the safety of the boiler and burner and performing the maintenance. While servicing a burner, the gas pressure switch diaphragm ruptured leaking natural gas into the room because it was not vented outside. I was unable to shut off the manual fuel valve and had to run outside and shut the main gas feed valve to the building. According to ASME CSD1, Section CF-150, *Valves shall be maintained and exercised to ensure valve remains operable without the use of tools.*

Types of Manual Gas Valves

Lubricated Plug Valve These valves were installed on some of the older gas boilers. They require lubrication once a year to be injected into the fitting on top of the valve. A dry plug cock could allow gas to leak from the valve. When injecting the grease into the valve, it should be exercised to make sure it is properly lubricated.

Ball Valve The ball valve is more commonly used now and should be tested regularly as well.

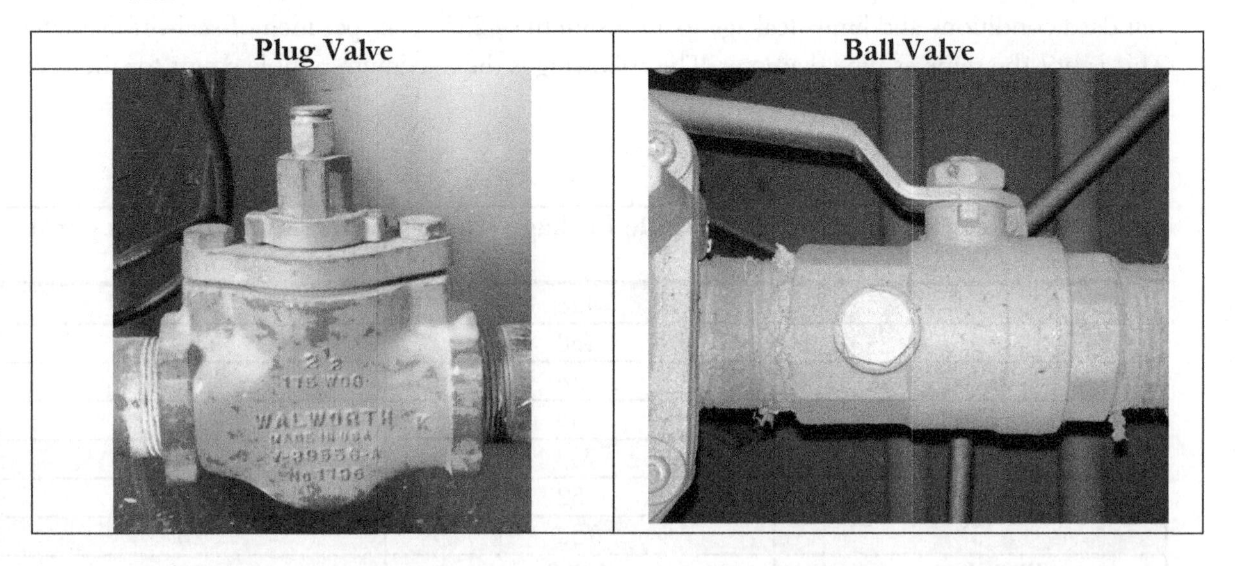

Plug Valve | **Ball Valve**

5 Firing Rate Valve This gas valve is connected to the modulating motor and is controlled by the modulating control. It will vary the gas to the burner as directed by the modulating or firing rate control.

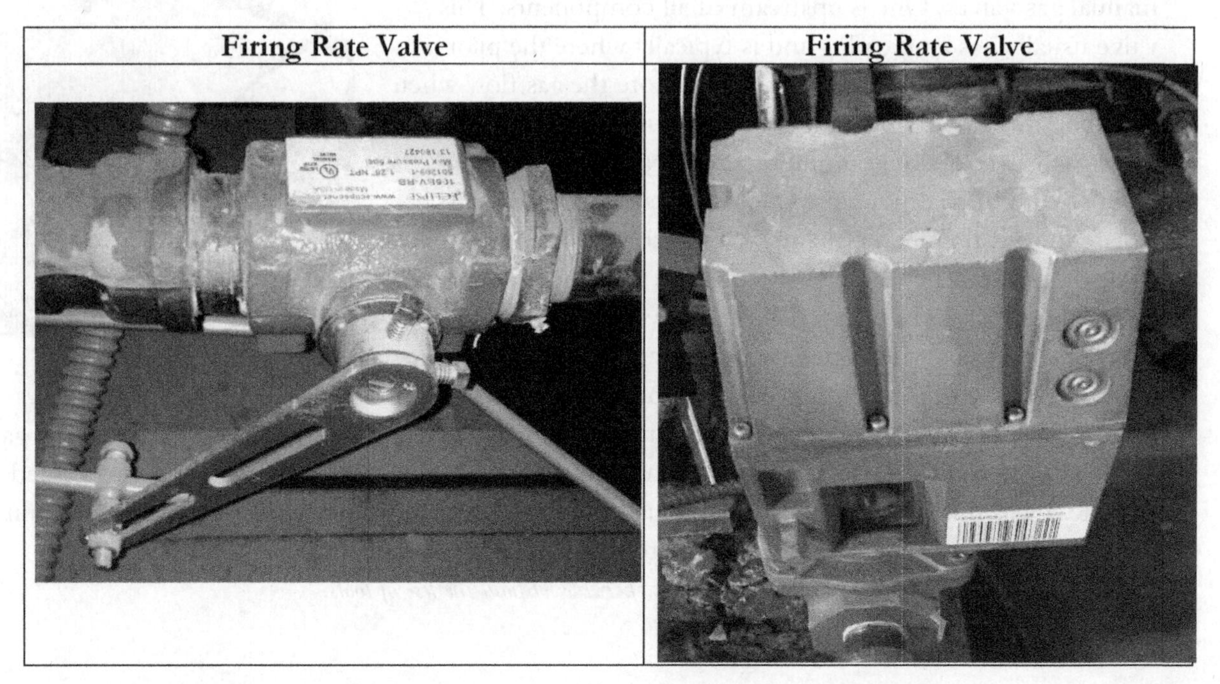

Firing Rate Valve | **Firing Rate Valve**

6 Modulating Motor This motor will vary the gas and/ or air to the burner as directed by the modulating or firing rate control and is controlled from the firing rate or modulating control. On the typical Honeywell modulating motors, there are three wires for control. R, W, B with R being the common. If you jump R-W, the motor will go clockwise or typically to low fire. If you jump R-B, the motor will drive counterclockwise or typically to high fire. According to the smart folks at T.F. Campbell in Pittsburgh, PA, you could also use a potentiometer to test the operation of the modulating motor. Please consult your Honeywell distributor for the potentiometer.

Modulating Motor	Modulating Motor

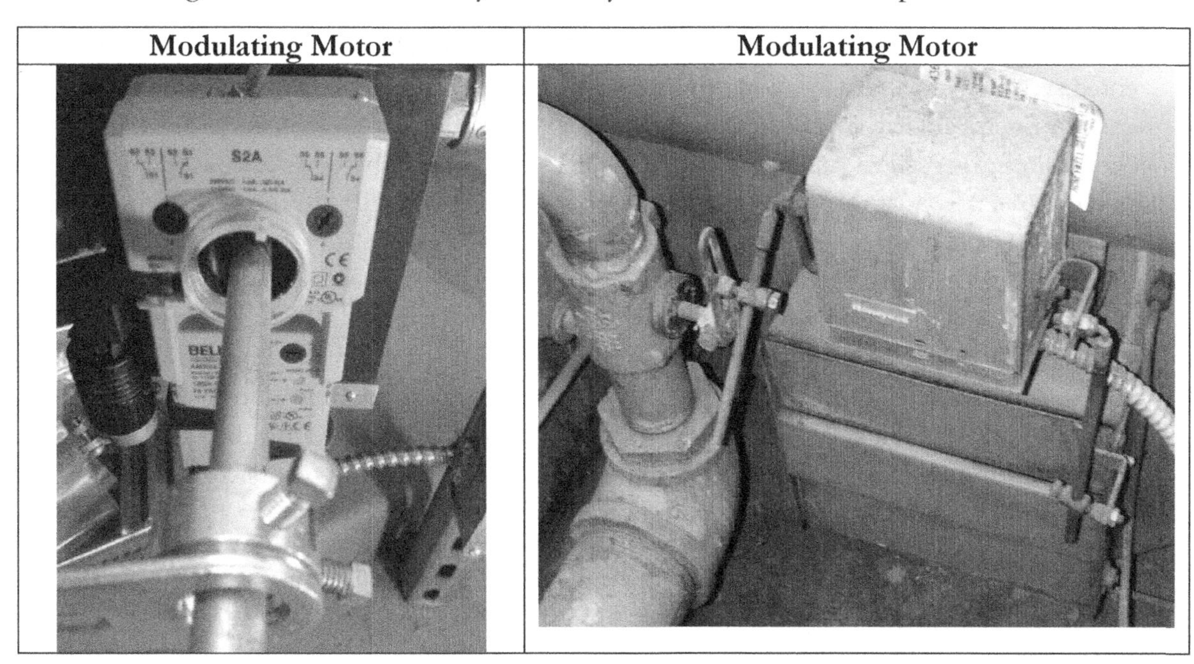

Low Fire Start Switch. Many of the modulating motors have an internal switch that is used to verify that burner is at a certain position. The most common switch is the low fire start switch. It will inform the flame safeguard that the burner is at low fire position so the pilot sequence can be started. In some instances, these have to be adjusted or even replaced.

Modulating or Firing Rate Control The modulating control will move the burner from low to high fire and anywhere in between to maintain the desired temperature or setpoint of the control. When adjusting the setpoint on these controls, they should be lower than the operating control setpoint. If the setting is higher than the operating control, it could cause short cycling of the burner. For example, if the operating control is set at 180 degrees F, the modulating control should be slightly lower. I would usually set it at around 175 degrees F. When the temperature at

the sensing bulb of the modulating control reaches 175, the burner should be at Low Fire. If the temperature or pressure drops, the burner will drive from low fire up to high fire to meet the setpoint. The modulating control is usually installed on the boiler but some of the older systems installed the modulating control on piping header to the building. The modulating control will vary the firing rate of the burner and that is called the proportioning or throttling range.

Are Gas Train Components Vented to the Outside? Many of the gas train components such as gas pressure regulators, gas valves and some gas pressure switches should be vented to the outside. Some old boilers piped the vents into the standing pilot or in the chimney. A common mistake is for the installer to vent all the components in a gas train in the same pipe as the Normally Open Vent Valve. According to UL 795, the normally open vent valve has to be vented separately to a safe location outdoors. No other vents can be combined with it. The normally open vent valve is located between the two electric gas valves in some gas trains. The vents from the regulators and gas pressure switches may be combined as long as the cross sectional area is at least the size of the largest vent opening plus 50% of the area of the additional vent pipes. Some new regulators will use a vent limiter that does not have to be vented to the outside. When servicing burners, one of the common service calls we see is the vent piping that get plugged, usually by spiders or bees. Verify that opening is clear and free. I suggest using a bug screen on the outlet. To see if the vent is plugged on a no heat call, verify that you have power and gas to the safety shutoff valve. You then disconnect the vent connection to the valves. If the burner fires, it is most likely the gas train vent. When cleaning the vent line, be cautious of using pressurized air as it could rupture the diaphragms on the other components connected to the vent. If using air, be sure to disconnect the vents from all other components and plug them.

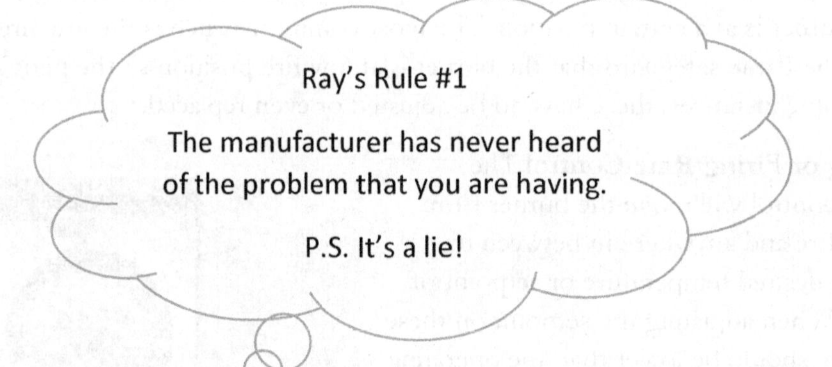

Sizing Gas Train Manifold Vent

Source: Philadelphia Gas Works

On older boilers, the vents used to be piped to the standing pilot as the flame would burn off any leaks. The newer boilers without a standing pilot require venting to the outside. A leaking 1/4" fitting could lose enough gas to fill a 10 foot by 10 foot x 10 foot high room with a combustible mixture within an hour. When combining common vents from gas train components, you need to see if the common pipe cross sectional area is large enough for all the components.

Let us assume that we will be venting two regulators with 3/4" vents and two gas pressure switches with 1/4" vents. We will get our cross sectional sizes from the chart below

3/4" regulators	0.533 X 2 = 1.066
1/4" gas pressures switches	0.104 X 2 = <u>0.208</u>
	Total 1.274

We would need a pipe size of 1 1/4" to combine all these vents.

- Pipe runs over 30 feet horizontal are not recommended and should be increased one pipe size for each 30 feet run.
- Normally open vent valves cannot be combined with other vents or other appliances.

Vent terminations should be:

- 4 feet below, 1 foot above, and 3 feet horizontally from windows, doors, and gravity air intakes. 3 feet above any forced air inlet within 10 feet horizontally.
- Vents should have a screened 90 facing down.

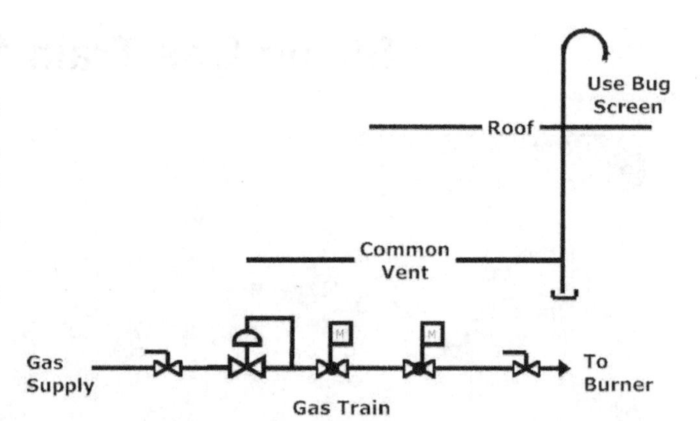

****ASME CSD1 Sizing: If your locale follows the ASME CSD1 code, Section CF-190 has a different formula for sizing the manifolded vent line. It requires that *"the manifolded line shall have a cross-sectional area not less than the area of the largest branch line directly piped to the manifolded line plus 50% of the additional cross sectional areas of the manifolded branch lines."***

Pipe Size	Inside Diameter	Inside Cross Sectional Area	50% of Inside Cross Sectional Area
1/8"	0.269	**0.057**	**0.03**
1/4"	0.364	**0.104**	**0.05**
1/2"	0.622	**0.304**	**0.15**
3/4"	0.824	**0.533**	**0.27**
1"	1.049	**0.864**	**0.43**
1 1/4"	1.38	**1.495**	**0.75**
1 1/2"	1.61	**2.036**	**1.02**
2"	2.07	**3.36**	**1.68**
2 1/2"	2.469	**4.788**	**2.39**
3"	3.068	**7.393**	**3.70**
4"	4.026	**12.73**	**6.37**
5"	5.047	**20.004**	**10.02**
6"	6.065	**28.89**	**14.45**

If you would like to combine the same size vents, this table can help.

Combining Gas Train Vents when using full cross sectional area.					
	Number of Vents				
Vent Pipe Size	2	3	4	5	6
¼"	½"	¾"	¾"	¾"	1"
½"	1"	1 ¼"	1 ¼"	1 1/2"	1 ½"
¾"	1 ¼"	1 ½"	2"	2"	2"

Pilot and Ignition

Pilot assembly The majority of flame failures on power burners occur during the pilot trial and verification period in the burner sequence. The pilot should be meticulously tested as this will save you service calls later in the heating season. The pilot tubing and pilot assembly sometimes fill with dirt and require cleaning. When testing the pilot, visually inspect for dirt. In severe instances, you may have to use small drill bits to clean out the orifice but verify that the drill bit is not too large. A welding supplier can sell you burner tip cleaners that you can use for cleaning a plugged burner orifice. In many instance, it is less expensive to just replace the pilot with a new one.

Pilot Assembly

Types of Pilots

Continuous or Standing Pilot The pilot flame burns constantly regardless if there is a call for heat.

Intermittent Pilot This pilot flame is ignited when there is a call for heat. It stays lit while the main flame is burning and will shut off when the call for heat ends.

Interrupted Pilot This pilot flame is ignited on a call for heat and only stays on until the main flame is ignited and verified.

Pilot Flame Verify that the flame is large enough to touch the flame sensing device. The flame sensing device could range from a thermocouple to a flame rod. Confirm that the combustion air or flame startup does not cause it to waver away from the flame sensing device. The pilot flame should envelope 3/8" to ½" of the tip of the thermocouple. The flame rod should be cleaned each year. Verify the pilot has a good flame and makes contact with the flame sensing device. In some boiler rooms, the wind from outside could blow out the pilot. To resolve the problem, you may have to install a metal shield in front of the boiler to divert the wind away from the pilot.

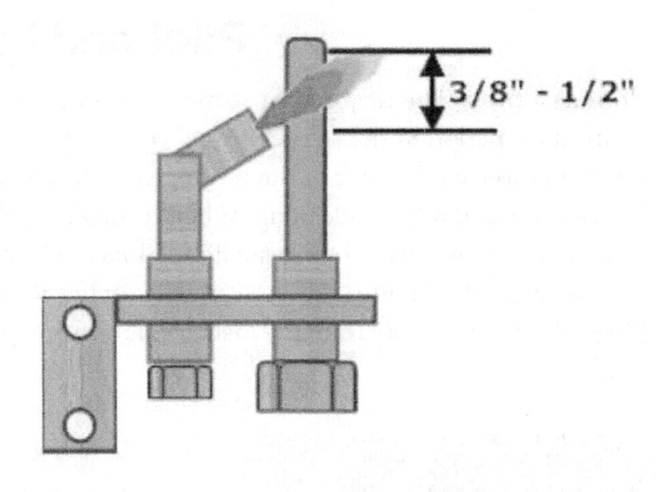

Pilot Adjustments During the burner service call, you may have to adjust the ignition electrode to get the proper spark required to ignite the pilot flame. There are certain gaps that the manufacturer requires between the ignition electrode and the ground source and the electrode and the burner nozzle. If the adjustment is incorrect, the pilot will not light. On many burners, you can look at the burner from the rear of the boiler to see if you see a spark. If the burner uses a flame rod to detect the flame, the gap on this will have to be set as well.

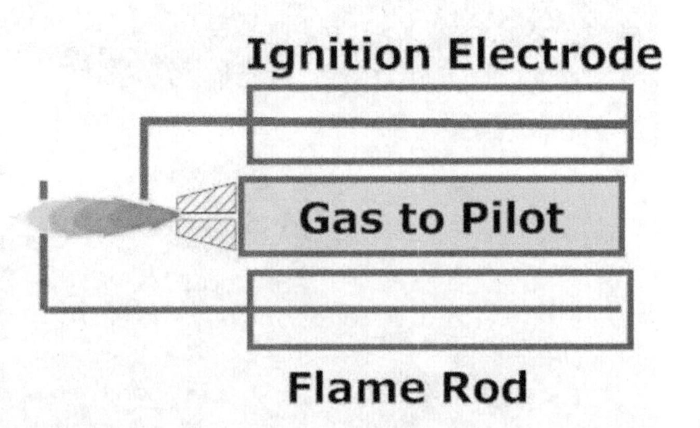

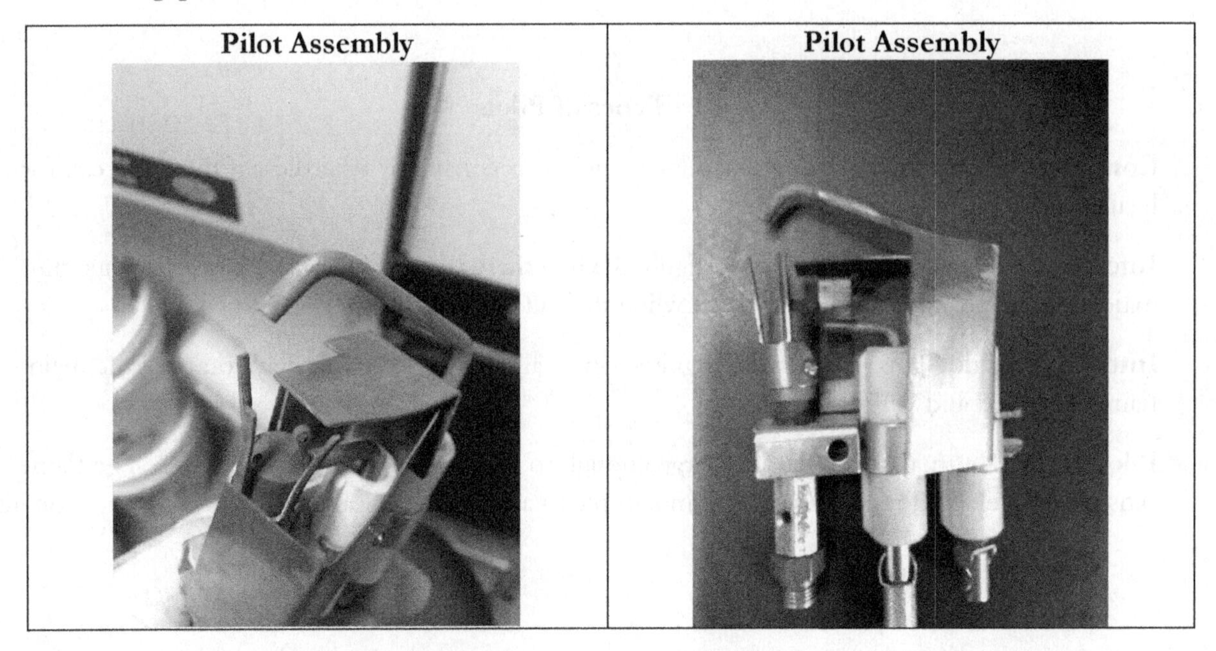

Pilot Assembly

Pilot Assembly

Pilot Tubing Verify that the tubing is not kinked or leaking. The pilot tubing is usually made of either aluminum or copper. There was a time when copper was not permitted to be used for natural gas. Please verify with your local codes to see if it is permitted.

Pilot Solenoid Valve The operation of the pilot solenoid valve is controlled by the flame safeguard control. It is usually a ¼" or 3/8" valve. To verify operation, place a gas pressure gauge or manometer downstream of the gas valve to verify opening of the valve.

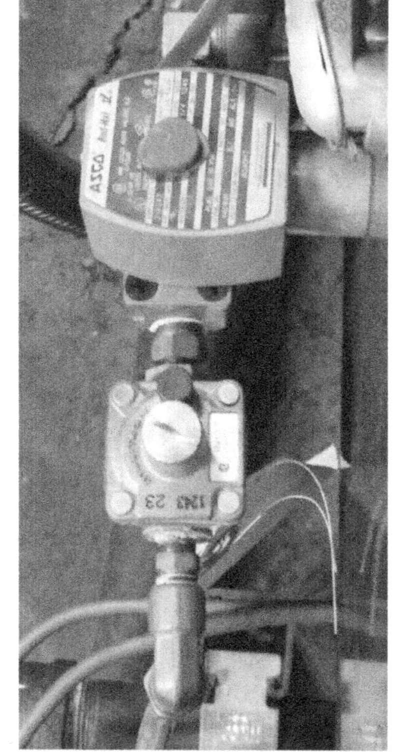

Pilot Gas Pressure Regulator The pilot gas pressure regulator assures proper gas pressure to the pilot assembly. To test the gas pressure regulator, you will need a gas pressure gauge or manometer downstream of the pilot regulator. Many burners have a fitting at the burner to allow testing of the pilot gas pressure. The burner manufacturer will specify the pressure they require to ignite the main flame. If you are having an issue with pilot flame failure, it is tempting to increase the gas pressure. In reality, this may cause more issues as it could push the pilot flame away from the flame sensing device. I have fixed many flame failures by lowering the gas pressure. I will monitor the flame signal when adjusting the pilot gas pressure to make sure it is within the proper range of the flame safeguard.

Ignition Electrode Most burner manufacturers use a spark to ignite the pilot flame, similar to a spark plug in a car. The flame safeguard will enable the transformer to send electricity to the ignition electrode. It is usually between 6,000 and 10,000 volts. If the cable is broken, it could short itself on the burner housing and not send the electricity to the electrode. The ignition electrode is usually a thin metal rod that is surrounded by white porcelain. The porcelain is an insulator that makes sure the electric current goes to the tip. When servicing the igniter, verify that the porcelain is intact and not cracked. From my experience, the porcelain will break if dropped. The burner manufacturer will tell you the gap that they require for the spark on the metal ends of the igniter. If the electrode is touching metal or too wide, there will be no spark.

Hot Surface Ignitor If the boiler uses a hot surface ignitor, you could check the continuity of the ignitor for cracks. Be careful with inspecting the ignitor as they could break easily once they have been exposed to flame.

Ignitor Pilot Assembly	Ignitor and flame rod

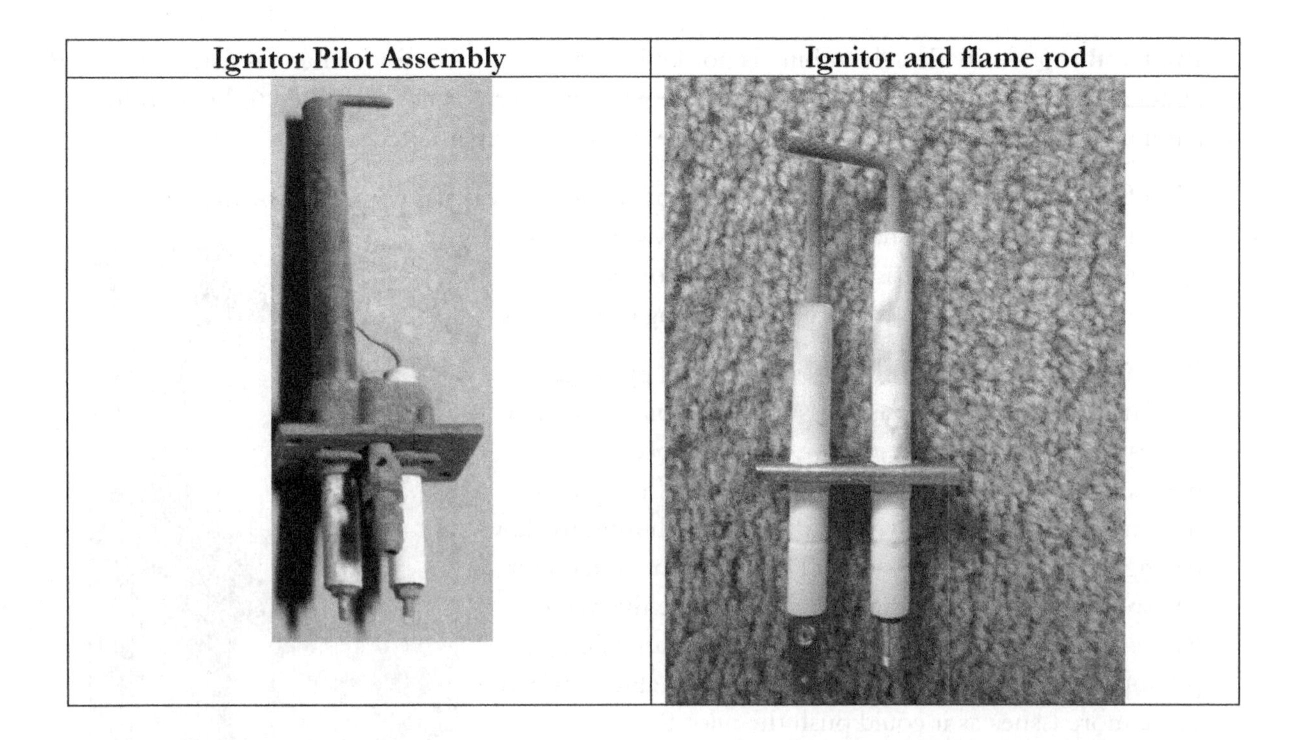

Transformer The transformer is used to deliver voltage to the ignition electrode. It will generate between 6,000 and 10,000 volts. Use caution when servicing this and make sure the power is off and confirm that the cable is intact and has no cracks or breaks. It will jolt you if you happen to touch a bare wire when it is powered. Verify that the ends of the wire have a firm grip on the connectors. If not, you could crimp them for a better grip. Be careful not to squeeze it too much. To test if the transformer is working, visually inspect that there is a spark on the ignition electrode when there is a call for heat. You should be able to see the spark from the sight glass of the boiler. If it cannot be seen, I have had to remove them from the burner and watch the spark. Remember to use extreme caution.

Ignition Temperature of Fuels			
Fuel	Degrees F	**Fuel**	Degrees F
Kerosene	500	**Coal**	850
Light Fuel Oil	600	**Propane**	875
Gasoline	735	**Methane**	1076
Butane-N	760	**Hydrogen**	1,095
Heavy Fuel Oil	765	**Natural Gas**	1,163
		Carbon Monoxide	1,170

Boiler Safety Controls

Low Water Cutoff The low water cutoff is the leading mechanical cause of boiler accidents. It should be tested on a regular basis and serviced at least twice per year. Consult the manufacturer or the building owner's insurance company to see the testing frequency they recommend. The low water cutoffs come in different types and configurations such as a float or probe. These are further described in the hydronic or steam section later.

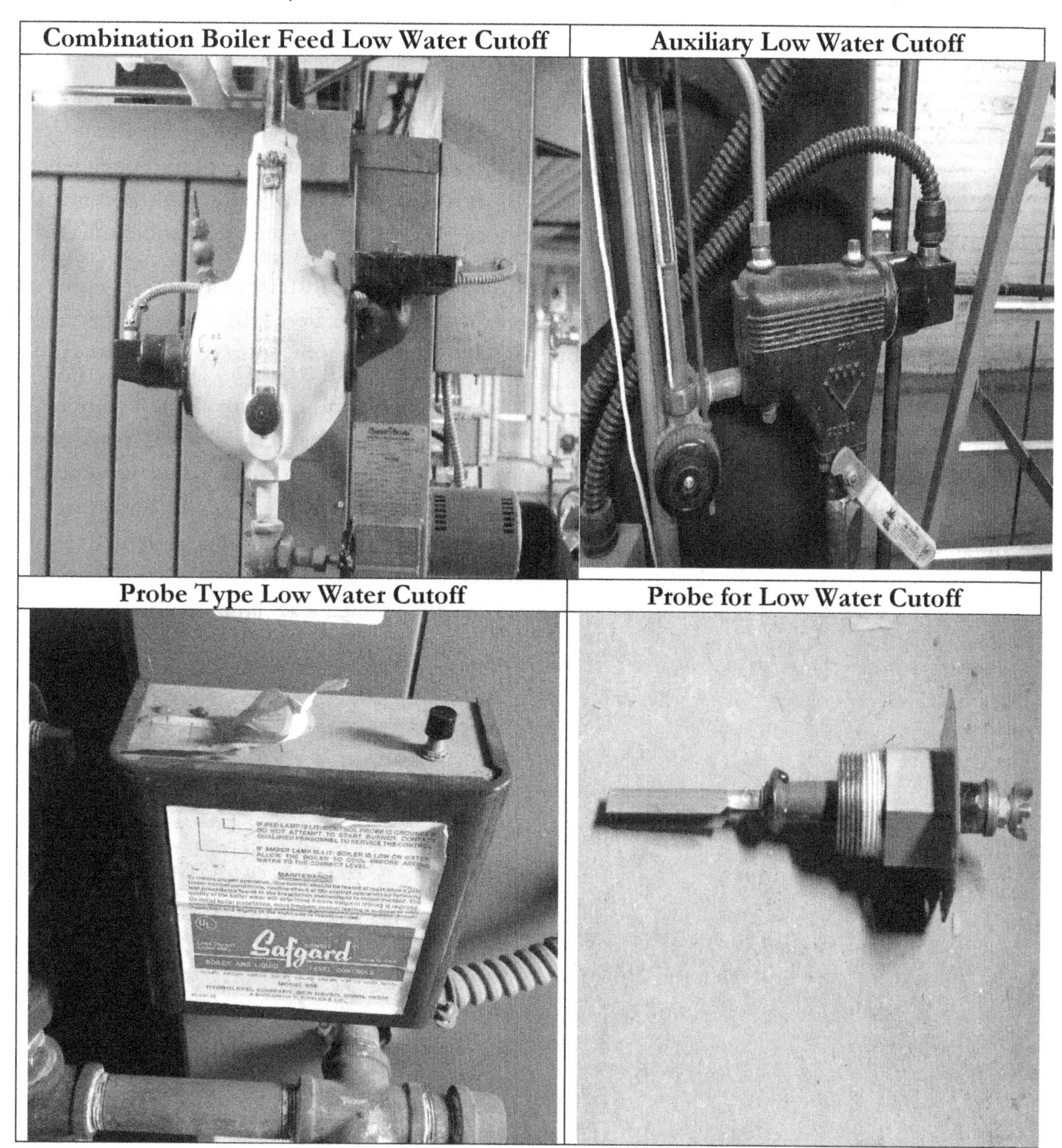

Combination Boiler Feed Low Water Cutoff	Auxiliary Low Water Cutoff
Probe Type Low Water Cutoff	Probe for Low Water Cutoff

Operating Control The boiler operating control is set at the desired temperature or pressure of the boiler. In most instances, the hydronic control will be set at 180 degrees F. For low pressure steam systems, most were originally designed for 2 pounds of steam pressure. When you are adjusting the setpoint, many of the controls are either subtractive or additive. For instance, a subtractive hydronic control that was set at 180 degrees F with a 5 degrees F differential would shut off the burner at 180 degrees F and would start the burner at 175 degrees F. If the control is an additive control, it will cycle at the differential above the setpoint. For example, if a steam boiler had an additive control and the setpoint was 2 psig and the differential was 2 psig, the boiler would cycle between 2 and 4 psig when there is a call for heat. Operating controls are an automatic reset control which means that it will operate between the two settings. To test these controls, note the setting on the control and observe the starting and stopping boiler temperature or pressure. These controls will make when the temperature or pressure drops and open when the temperature or pressure is above the setpoint. It is rare to have the setting on the control to match the actual on and off temperature or pressure. As a First Born guy, that drives me crazy.

Differential The differential is the variance between the on and off temperature or pressure of the control. The wider the differential, the longer the burner operates. I prefer a wider differential as it increases the efficiency of the boiler but you have a wider temperature or pressure swing.

Limit Control The limit control is similar to the operating control but with a couple differences. The setting for the limit control is higher than the setting of the operating control. In addition, it is a manual reset control. It means that if the temperature or pressure raises to the setpoint of the limit control, the control will shut off the burner and will not allow operation of the burner until the temperature or pressure is below the setting and the manual rest button is pushed. To test these, you will have to either lower the setting and compare it to the actual temperature or pressure or raise the operating control setpoint above the manual reset limit control.

Never try repairing the safety controls, which include the operating, limit, low water cut-off or flame safeguard controls. You could void the warranty and make yourself personally responsible for any problems.

Did you know that the heat released from a radiator is only 40% radiation and 60% convection?

Other Items to Inspect

Boiler Isolation or Shutoff Valves should be exercised to see whether they will operate properly. Operation should be verified in case you need to isolate the system from the boiler. For example, if you have a repair on the boiler that requires draining the boiler, isolation valves will allow you to keep the water in the system without draining the entire building. This will save the time it takes to refill and vent the system and the chemicals lost for the water treatment. Do not shut off valves when pumps are operating as it could "Dead Head" the pump and damage it. "Dead Head" means that the pump will be working against a closed valve.

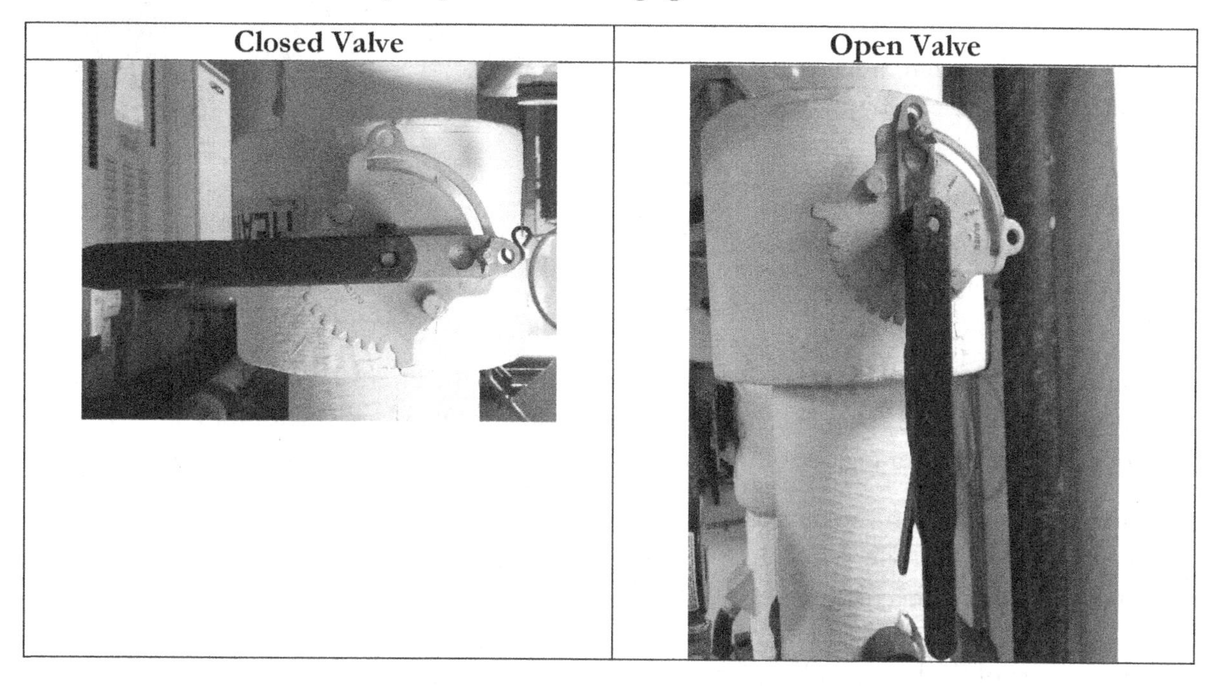

Closed Valve | **Open Valve**

Steam Boiler Valves installed on steam boilers make it easier to test the steam pressure controls. You can close the valves and watch the steam pressure build and note the steam pressure where the pressure controls operate. You can then open the valve and see the steam pressure drop. Use extreme caution when opening the steam valve into a cold pipe as condensate induced water hammer could occur. This happens when steam is surrounded by cold condensate and causes rapid condensation. It will generate pressures well over 1,000 Psig.

Gate Valves I prefer a rising stem gate valve or a ball valve for isolating boilers so you will know immediately if the valve is open or closed. If the boiler shutoff is a non-rising stem gate valve, you have no idea if the valve is open or closed by visual inspection. When I have one of these, I will open the valve all the way and then close it ½ turn to limit the chance of the valve being frozen in place.

Gate Valve

Boiler Base Check for cracks or signs of broken parts. Verify that the insulation or refractory is intact.

Confined Space If you have to enter the boiler to check the internal components, it should be considered a confined space and follow your company or OSHA guidelines for entry. This should include lockout tag out of the electrical and closing the manual fuel valves to the burner.

Boiler Internals On some service calls, you may have to open and inspect the internal components of the boilers. In many states, the boiler inspector for the municipality will inspect the internal components of the boiler looking for areas that are not safe. Some boilers such as cast iron sectional and fire tube boilers are difficult to access the water side of the unit. Fire tube boilers typically have hand hole and man hole openings to allow inspection. Some water tube boilers allow access to both the water side and fireside. If you are going to open the boilers, you should purchase the gaskets from the manufacturer to allow you to close the boilers.

Manhole Opening

When opening the boiler access doors, they will be extremely heavy. It is not a one person job. I have also found that some boilers with hinged access doors may not line back up due to the play inside the aging door hinges. In addition, many of the doors contain refractory which could crack or break if treated roughly.

Burner Maintenance

Main Burner The main burner should be checked a minimum of twice per year.

Atmospheric Burner If the burner consists of ribbon burners, they should be inspected for holes and rust, other than the factory made openings. Check for blockages inside the burner tubes and flakes of rust on the top of the burner. Spiders will sometimes build webs inside the tubes. They should be cleaned and visually inspected to see if the flame is even across the burner. When watching the flame, verify that it goes straight up and does not impinge on the side of the boiler walls and the flame is not be too high as to impinge on the boiler surfaces. If the flame is pulled up into the boiler tubes, it may be too high. If it is being pulled high into the boiler, check the draft of the stack. Excessive draft could pull the flames into the boiler and lower the efficiency of the unit. Check the gas pressure to make sure it is at the proper pressure. Older atmospheric burners typically used about 3 1/2" W.C. gas pressure.

Atmospheric Boiler

Power Burner Power burners are a bit more difficult to inspect as they are installed inside the boiler and usually surrounded by refractory. You may have to enter through a man door to inspect the inside of the boiler. I have seen people using bore scope cameras or mirrors to inspect the burner head. Look for damage to the head and warping or distortions. While looking at the burner head, inspect the surrounding refractory. Look for missing pieces or cracks. Make sure the burner orifices are intact and not blocked.

Burner mounting plate Check the burner mounting plate as there could be leaks. The rope gasket sometimes dries out and will allow air and possibly combustible gases to leak out into the boiler room Be sure the mounting is secure to the boiler. On older burners, they used asbestos rope to seal the burners. Asbestos is known to be dangerous if inhaled.

Air Fuel Interlock This control is usually mounted on the burner or inside the burner control and its job is to verify that the blower wheel is operating before any fuel is introduced. It is usually a differential pressure switch or an air flapper switch. To test this control, I will check the continuity of the air switch when the blower is running. Verify that the movement is free and does not bind throughout the range of motion. If it is differential pressure switch, be sure the sample tubes are open and clear.

Differential Pressure Switch	Air Flapper Switch

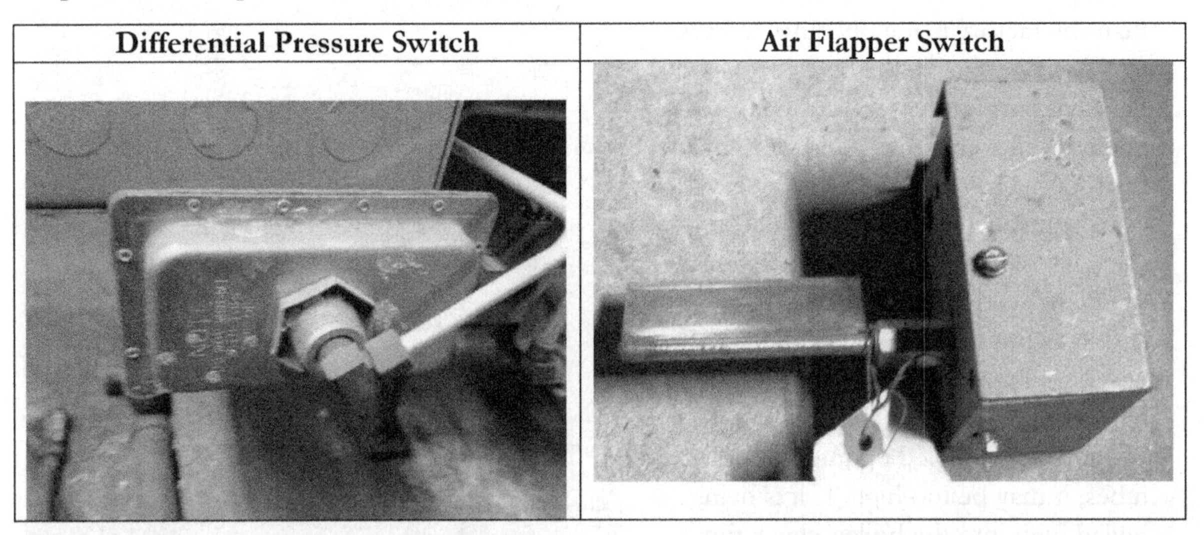

Blower Wheel The blower wheel inside the power burner is connected to the blower motor and introduces air to mix with the fuel. It is usually operating at over 1,700 revolutions per minute and sometimes over 3,000 RPM. When servicing a power burner, check the cleanliness and balance of the wheel. If the wheel is operated out of balance, it could be damaged. If you clean the blower wheel, make sure it is completely clean it as dirt on parts of the wheel could affect the balance. Use caution as to not disturb the balancing weights on the blower wheel. Be sure that the wheel does not rub the housing. Verify that the locking screws that hold the wheel to the motor shaft are tight and secure. Another consideration when cleaning a blower wheel is that the fuel to air ratio should be checked as you are now introducing more air and that will affect the combustion efficiency of the boiler.

Blower Motor Most blower motors operate at over 1,700 revolutions per minute in a hot moist environment for several thousand hours per year. This could lead to excessive wear. When servicing the motor, listen when it starts and stops as worn bearings sound raspy or like two sheets of sandpaper being rubbed together. Check for excessive vibration as this could indicate worn bearings. Some older motors had oil ports or grease fittings. A few drops of oil or grease will extend the life of the motor. Clean the air holes in the motor so that it will not over heat. Do not stick anything into the vent holes as it could damage the motor windings. I like using a soft tooth brush and a shop vacuum to clean dirty air holes. Verify that the blower motor rotation is correct. This should be done prior to any fuel valve opened.

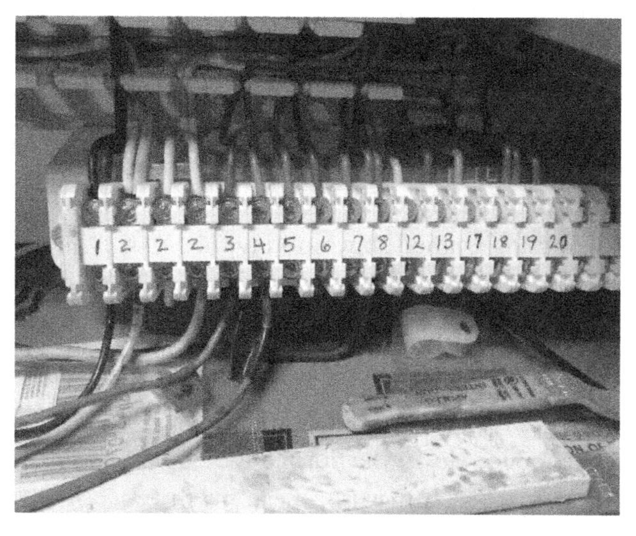

Electrical Terminals Inside the burner control panel are wiring terminal strips. Check the tightness of all the wiring connections. I find loose connections on almost every burner service call. The high speed of the blower and vibration causes the wiring connections to loosen. A loose wire could cause intermittent problems or even short a component. Be sure to shut off power to the burner before reaching inside. To avoid the chance of a shock or shorting a wire, I will use a screwdriver that is covered with electrical tape to the tip as the spacing between the terminals is tight.

Refractory is used to reflect heat back into the boiler and protect the internal metal surfaces inside the boiler. Although I have seen some much older, the life expectancy of refractory is only about 10 years. When refractory is exposed to flame, it becomes brittle and could flake or crack if touched. Always use respiratory protective equipment to remove or replace the refractory. If you do have to replace some of the refractory inside the boiler, please read the instructions and consult the MSDS sheets as old refractory could contain dangerous materials, such as silica or asbestos.

In most instances, refractory has to be cured and air-dried before heat is applied. If flame is applied to refractory when it is wet, it could cause the refractory to fall apart. The moisture inside the refractory could turn to steam and explode out causing the metal to be exposed to the flame. Metal exposed to flame causes embrittlement of the metal, weakening it. You could actually break it apart with your hand. It makes you feel like Superman. My first job servicing boilers involved me laying inside a warm boiler pounding refractory around a new burner head to protect the burner housing. It was not a good day.

Linkages The linkages are used to assure the proper fuel to air ratio throughout the firing range of the burner. Make sure the screws holding the rods in place are secure and the swivel connectors are not worn. I have seen the ball fall out of the swivel connector and the damper slammed shut, sooting the boiler.

Burner Characterized Linkages

Clean Up When Done This was taught to us as children and it is a good idea when servicing boilers. I like wiping down the burner and boiler jacket. It may not mean much but it shows that you respect the client and their property. The client will always take notice although they may not say anything.

Don't Kill The Messenger I like to meet with the building owner when the service call is complete to explain my findings. If there is a dangerous condition, I follow up with a letter explaining the findings.

When I finished the boiler preseason check, I found that the internal float on the low water cutoff would hang up and simulate a boiler filled with water when it was actually empty, which is dangerous in case you were wondering. I reviewed my findings with the maintenance director for the school and explained the dangers of a defective low water cutoff. The maintenance director for the facility reminded me of the professor from the movie, Back to the Future as his hair went in every direction and was never combed. His shirt would often be cross buttoned making it appear crooked. He always wore open rubber galoshes, regardless of the weather. He gave us permission to replace the low water cutoff. When I arrived a few days later, a meek young lady caught me in the hall and asked if she could talk to me. She told me that she was the kindergarten teacher and her classroom was right next to the boiler room. She then told me that she has not slept since I was there a few days earlier. I was flattered and told her that I was married. She looked at me like I had the plague and said, "No, that's not it. The Maintenance Director told all the teachers in the cafeteria about the defective control on the boiler." He also said that the boiler could have exploded at any time, killing her and all the kids as it would rocket through her classroom. Her heart raced at each noise she heard and she could not concentrate. "Why would they have something that dangerous in a school?" she asked. I explained that the boiler was not operating since I found the defective control as I had shut it off. I also informed her that the boiler has redundant safety controls. This look of relief crossed her face and she forced a smile and thanked me. Be careful how you tell the customer what is wrong.

Burner Service Tools

When servicing burners, there are several items that I like to carry. I keep my burner service tools in a separate tool box so that I have everything I need in one place. The following items are in my burner service box:

Electrical Extension Cord I always bring an extension cord with several female outlets because either my analyzer or drill will need recharged.

Electric Meter When using your meter for burner service, make sure it can read the range of the flame safeguard controls.

Drill and Bits The combustion analyzer probe requires a hole in the flue so I keep a drill bit set and cordless drill in my start up box. I also keep a single drill bit large enough to fit the analyzer probe in each combustion analyzer case.

Electrical Jumpers You may need to use electrical jumpers to fool the burner into a call for heat as many buildings have it connected to the building's control system. Jumpers are just for testing and should never be left on a boiler control.

Battery Operated Screwdriver I use these for testing the steam pressure controls as it makes it easier to turn them up or down.

Carbon Monoxide Detector It is always a good idea to test the boiler room for carbon monoxide so that you know it is a safe environment to work.

Trouble Light The boiler always seems to be in the darkest places so I will take in a trouble light.

Knee Pads When working on burners, you will be on your knees. This can really make you sore and could affect your concentration. I like taking either knee pads or something soft to kneel on.

Manometer Burner service technicians should carry a manometer to measure the gas pressure to the burner and the boiler draft. This could be the older manual ones filled with fluid while I prefer the digital manometers as they are easier to read for me. The manometer could also be used for testing draft.

Gas Pressure Gauges In addition to manometers, I have an assortment of gas pressure gauges. While these may not be as accurate as a manometer, they will allow you to trouble shoot valves to see if they open.

Batteries I keep an assortment of batteries in my tool box in case the batteries die in the middle of a service call.

Nipples and Fittings I like having an assortment of ¼" and 1/8" nipples and fittings. I keep some longer nipples to reach the gas pressure test fittings that are difficult to access. The box also contains bushings or reducers.

Allen and Torx Wrenches You may need Allen wrenches to access the test plugs. Torx wrenches are used to adjust some of the European styles burners.

Pipe Dope / Teflon Tape I keep a small can of Non Teflon pipe dope as well as some Teflon Tape to apply to the threads to reinstall the plugs for the gas pressure test ports. Be sure that the burner manufacturer allows the use of Teflon.

Orifices Some burners use orifices in the gas train to limit the gas pressure to the burner. If you are working on burners like that, you should have an extra set in case you need to change orifices and the burner firing rate.

Screws or Tape for Combustion Test Hole. When I am finished with the combustion tests, I will plug the test hole with either a lag screw or cover it with metal tape. Some new boilers have a positive flue and check with the manufacturer as to how they recommend sealing the combustion testing hole because the flue gases could be pushed into the boiler room.

Draft Gauge Verify the boiler draft is within the parameters required by the manufacturers. If the draft is excessive, it could lower the efficiency of the boiler. On Category 1 boilers, the draft should be slightly negative, usually about -0.05" w.c. Some analyzers have an internal draft tester.

Electronic Manometer

Combustion analyzer Many techs think they can simply eyeball the flame to adjust the fuel to air ratio. If you do, the adjustments will not be correct. A calibrated combustion analyzer is the only true way to properly adjust the ratio. I like keeping a spare analyzer with me in case the primary one breaks down. I had an oxygen cell fail during a service call four hours from home.

Nice to Have Tools:

Honeywell Digital Readout The Honeywell 7800 series flame safeguard is very popular in commercial burners. I like the using the digital display that is available through the Honeywell distributors. It allows you to see the history of the flame failures which is helpful for troubleshooting. It also allows you to monitor the flame signal of both the pilot and the main flame.

Purge Timer When trouble shooting a power burner, the time it takes the purge timer to run seems like an eternity. The purge timer is used to purge the boiler and flue passages of any combustible materials. It could be anywhere from 30 seconds to several minutes. A tech I know a keeps a ten second purge timer for the Honeywell 7800 series flame safeguard in his startup kit to make the process quicker. He will use this only when trying to troubleshoot a pilot issue.

Flame Safeguard

Flame Safeguard Without a doubt, the most important safety device on a boiler is the flame safeguard control. The flame safeguard could be anything from a thermocouple that costs a few dollars to a microprocessor unit that costs several thousand dollars.

Thermocouple Although standing pilots are rare anymore, we still see them on older boilers. The thermocouple will generate voltage in the presence of flame. If working properly, the thermocouple will generate about 30 mV or millivolts. I like to change the thermocouple yearly to avoid no heat calls and will leave a spare on the jobsite for those Friday night calls. The thermocouple will be connected to either the gas valve or one of the older type pilot safety valve or pilot safety switch, such as a Baso switch. Visually inspect the thermocouple. If the end is split or spalled, it should be replaced. There are some testers available at the wholesale houses to check thermocouples. I have found that it is less expensive to simply change the thermocouples yearly. When testing the thermocouple, the main flame should ignite within 4 seconds without rollout, according to Honeywell sheet 60-2087-4.

Thermocouple connected to Baso Switch	Pilot Safety or Baso Switch

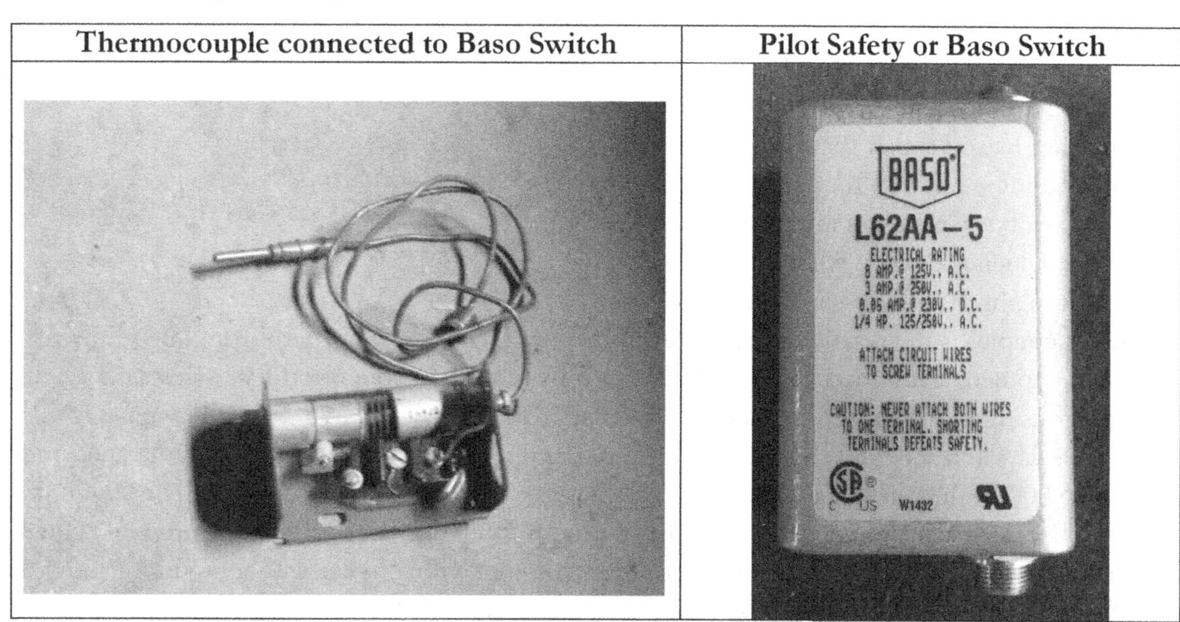

Did you know that 75% of the electricity used to power home electronics is consumed while the products are turned off?

Pilot Safety Switch or Valve Boilers with thermocouples commonly were attached to a device called a Pilot Safety Switch or commonly called the Baso switch or they could be connected to a pilot safety valve. The thermocouple would send the flame signal or lack thereof to this switch and it would control the main gas valve. To test these, you can get a tester from your flame safeguard distributor that allows you to check operation. A simple way to test them is to see if they trip when the flame is gone from the thermocouple or you can unscrew the thermocouple from the switch to see if it shuts off the gas.

Pilot Safety Valve

Many people ask me why we no longer see thermocouples on the job as they were easy to repair and inexpensive for the customer. The main reason is that the codes do not allow standing pilots as they are wasteful. Most new burners use an interrupted type of pilot which is only on when the burner is trying to light. Another issue with thermocouples is the reaction speed. A thermocouple may take anywhere from one to three minutes to detect and react to a loss of flame. This could be enough to fill a boiler with a dangerous mixture of combustible gases and cause an explosion. Most of the new flame safeguards will react within a few seconds, making it safer. For example, if the flame goes out on a 1,000,000 Btuh boiler with a thermocouple and the thermocouple response time is 2 minutes, the boiler will have about 16 cubic feet of gas and mixed with the combustion air would equal about 240 cubic feet of explosive mixture. Contrast that with a boiler with a modern flame safeguard with a 4 second response time. It would allow slightly more than 1 cubic foot of gas and the explosive mixture of air and gas would be about 16 cubic feet, about 94% less.

On a call for heat, the flame safeguard will start the burner fan. The blower operation is verified with a sensing device, usually a pressure switch. If the burner is a modulating one, the burner air intake damper will open. The burner enters what is called "Pre-Purge." The fan will operate for anywhere from 30 seconds to several minutes. This is to "purge" the boiler combustion chamber of any unburnt fuel. The purge timing will be long enough to provide four full air changes in the combustion chamber. After the pre-purge, the pilot will light. The flame safeguard will check the pilot flame to make sure that it is safe and stabile. If the burner is a modulating one, it will drive to a low fire position before igniting the pilot.

Older Mechanical Flame Safeguard Meter connects here with Honeywell adaptor	Microprocessor Flame Safeguard Where to connect meter

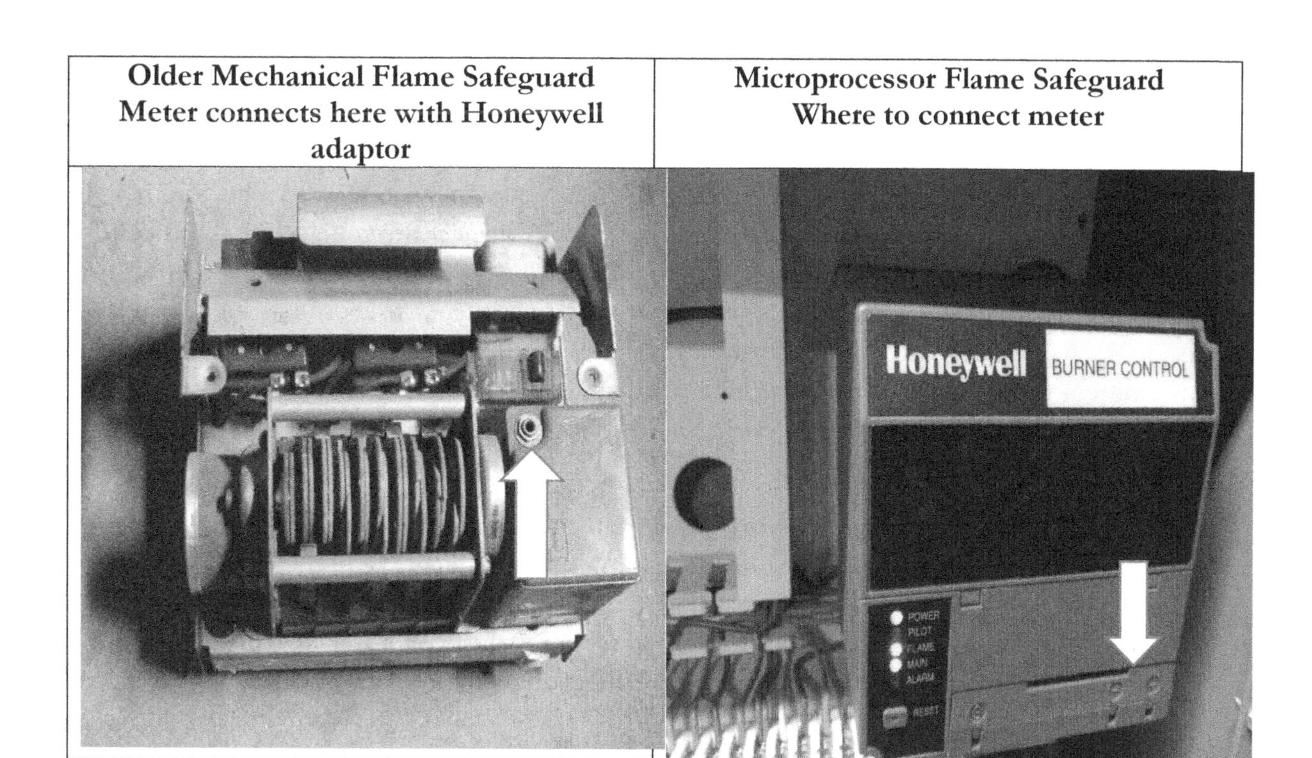

At that point, the main gas valve will open. The flame safeguard will continually monitor the flame until the call for heat has ended. Some older burners used a "Post-Purge" to void the combustion chamber of any unburnt fuel after the call for heat. This is wasteful as it will take heat from the boiler and exhaust it up the stack. The best way to check the operation of a flame safeguard is with a meter. On Page 68, I have a table showing the suggested readings for different flame safeguards.

Melted Flame Safeguard	Honeywell 7800 series digital display

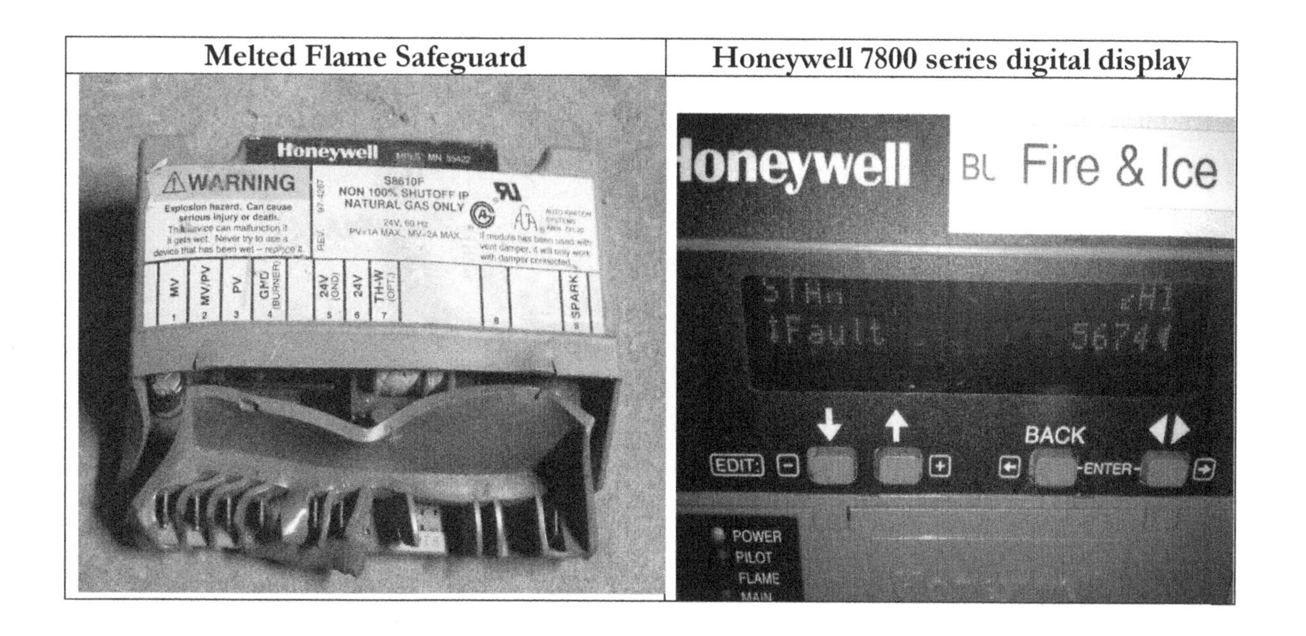

Common Flame Safeguard Maintenance Use caution when servicing flame safeguards as they may still be powered even though the burner is off. Check for voltage before servicing. Each flame safeguard manufacturer has their own set of maintenance and safety checks they recommend. I like to keep a copy of the maintenance requirement to refer to while on the job site. I pull the flame safeguard from the base and visually inspect the contacts on the back of the control. Once the power is verified off, I check the tightness of the screws for the flame safeguard. Since the burner vibrates, these screws have a tendency to loosen. A loose connection inside the flame safeguard can drive you nuts. That is the first thing I do when I have an intermittent flame failure. When inspecting the flame safeguard, look for black trails that could signify shorting wires. You want to verify that the contacts are in good condition on the rear of the flame safeguard as well.

Replacing the Flame Safeguard Many are tempted to use rebuilt flame safeguards as they are much less expensive than new or factory rebuilt ones. Some states like mine prohibit the use of rebuilt flame safeguards. I was on a job where the customer purchased a rebuilt flame safeguard. It allowed the main gas valve to open without verifying that the pilot was lit. This could have been very dangerous. I informed the customer of this and shut down the boiler.

Typical Sequence of Operation for Power Burners

1. Call for heat.
2. Verify all safety and operating controls are closed.
3. Burner starts blower motor.
4. Airflow switch is made.
5. Burner starts pre-purge period which could last up to 2 minutes.
6. Electricity is sent to ignition electrode and pilot solenoid valve is energized.
7. Trial for ignition starts and should be from 4-10 seconds long.
8. Once ignition is proven, power is supplied to main fuel valves.
9. Flame safeguard will supervise ignition of main flame.
10. Call for heat ends.

Ultraviolet Flame Sensor	

Flame Sensing

Another consideration with flame safeguards is the flame sensing types. There are commonly four flame sensing devices. They include flame rectification or more commonly known as a flame rod, photo cell, ultraviolet, and infrared. Flame rectification is the most common flame sensing device and it uses a theory that a flame will actually conduct electricity. I wonder how many singed eyebrows there were when they tried to test that theory. A small electrical charge will be sent to the flame rod. If a flame is present, the current will travel through the flame to ground on the burner. If the ground is sensed, the main gas valve will open.

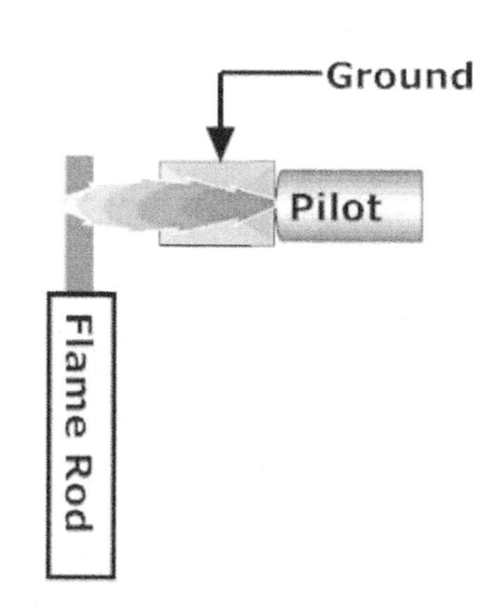

Four to One Ratio When servicing burners with rectification or more commonly called a flame rod, the four to one ratio should be followed. The grounding surface that is in contact with the pilot flame should have an area that is four times the area of the flame rod that is in contact with the flame. In some instances, you may get a higher flame signal when less of the rod comes in contact with the flame.

The **photo cell**, common for fuel oil burners, looks for the visible light of a flame. When light is present, the sensor will emit electrons and tell the flame safeguard that a flame is present.

Infrared flame sensors can be used on either gas or fuel oil. It senses the infrared radiation that is present in a flame. The sensor enclosure is usually a red color. It can sometimes be fooled by hot refractory inside a boiler.

Ultraviolet flame sensing tests for the ultraviolet waves that are present inside a flame. The UV sensor will not respond to any visible light. The sensor enclosure is usually a purple color. It can be fooled by hot refractory, spark ignition, gas laser, sun lamps, germicidal lights and a bright flashlight.

Typical Flame Safeguard Signals	
μA = Micro Amp	***VDC = Volts DC***
Fireye	
Model	**Average Flame Signal**
UVM	4.0-5.5 VDC
TFM	14-17 VDC
D-10/20/30	16-25 VDC
E-100/ 110	20-80 VDC
E-100/E110 with EPD Programmer	4-10 VDC
M Series II	4-6 VDC
Micro M Series	4-10 VDC
Micro M Series w Display	20-80 VDC
Honeywell	
Model	**Average Flame Signal**
RA890	2-6 μA DC
R4795	2-6 μA DC
R7795	2-6 μA DC
R4140	2-6 μA DC
R4150	2-6 μA DC
BC7000	2-6 μA DC
RM7890	1.25-5 VDC
RM7895	1.25 VDC
RM7840	1.25 VDC
RM7800	1.25 VDC
S8600	1-5 μA DC

Typical Flame Safeguard Signals Siemens	
μA = Micro Amp	***VDC = Volts DC***
Gas Burner Controls	
Model	**Flame Signal**
LFL with UV sensor QRA **Minimum 70 uA DC**	100-450 μA DC Typical
LFL with Flame Rod **Minimum 6 uA DC**	20-100 μA DC
Oil Burner Controls	
Model	**Flame Signal**
LAL1 with photoresistive detector, QRB1	95-160 μA DC
LAL1 with blue-flame detector, QRC1	80-130μA DC
LAL2/LAL3 with photoresistive detector, QRB1	8-35 μA DC
LAL2/LAl3 with selenium photocell detector, RAR	6.5-30 μA DC
LAL4 with photoresistive detector, QRB1	95-160 μA DC
LAL4 with blue-flame detector, QRC1	80-130 μA DC

Pipe Insulation

Uninsulated pipes are usually a bad thing in boiler rooms. In addition to increasing the heat in the boiler room, it lowers the efficiency and could affect the operation of the system. An uninsulated steam pipe will cause the steam to condense. This cools the steam, reducing the efficiency and increases the maintenance of the system. The International Energy Conservation Code requires pipe insulation on all projects. The following are their recommended insulation thicknesses.

Minimum Piping Insulation 2009 International Energy Conservation Code 503.2.8

Heating Medium	Pipe 1 ½" and below	Pipe > than 1½"
Steam	1 ½"	3"
Hot Water	1"	2"

Insulate Condensate Pipes? If you want to get more life from the condensate piping, consider insulating it as insulation reduces the carbon dioxide inside the pipe which becomes carbonic acid when mixed with water.

Could Insulation Cause Pipe Corrosion? According to an article entitled, " Inspection Techniques for Detecting Corrosion Under Insulation" by Michael Twomey, corrosion under insulation is caused by the water trapped inside the insulation. The insulation can hold the water like a sponge and keep it in contact with the metal surface. The water can come from rain water, leaks, sweating from temperature cycling, or low temperature operation such as refrigeration units. Corrosion becomes a significant concern in steel at temperatures between 32°F (0° C) and 300°F (149° C) and is most severe at about 200° F (93° C). Corrosion rarely occurs when operating temperatures are constant above 300°F (149° C). A concern when insulating pipes is corrosion that occurs under the insulation. The insulation could hide and sometimes precipitate the occurrences of corrosion.

Some of the areas that may be susceptible to corrosion under the insulation is

- Pipes that are exposed to mist overspray from cooling water towers.
- Pipes exposed to steam vents.
- Pipes that experience frequent condensation and re-evaporation of atmospheric moisture.
- Carbon steel piping systems that normally operate in-service above 250° F (120° C) but are intermittent service.

- Deadlegs and attachments that protrude from insulated piping and operate at a temperature different than the active line.
- Vibrating piping systems that have a tendency to inflict damage to insulation jacketing providing a path for water ingress.
- Steam traced piping systems that may experience tracing leaks, especially at the tubing fittings beneath the insulation.
- Piping systems with deteriorated coatings and/or wrappings.
- Locations where insulation plugs have been removed to permit thickness measurements on insulated piping should receive particular attentions.

Uninsulated Steam Pipe

Mold Be careful if using fiberglass insulation when exposed to chilled water in a boiler room. I have seen black mold growing in the insulation due to the pipe sweating. Black mold is very dangerous and could cause health issues if inhaled.

Relief or Safety Valves

A series of malfunctions occurred in the boiler room that evening. The steel compression tank for the hydronic loop flooded, leaving no room for expansion. Water will expand at 3% of its volume when heated from room temperature to 180°F. When the burner fired, the expansion of the water increased the system pressure within the boiler. The malfunctioning operating control did not shut off the burner at the set point which caused the relief valve to open. The brass relief valve discharge was installed with copper tubing piped solid to a 90 degree ell on the floor and the tubing further extended to the floor drain. The combination of hot water and steam from the boiler caused the discharge copper tubing to expand, using the relief valve as a fulcrum. The expansion of the copper discharge tubing pressing against the floor was enough to crack the brass relief valve, flooding the boiler room. The damage was not discovered until the next morning, several hours after the leak occurred. Thousands of dollars in damage was sustained and luckily no one was injured.

Relief Valve

Each boiler requires some sort of pressure relieving device. They are referred to as either a safety, relief, or safety relief valve. While these names are often thought of as interchangeable, there are subtle differences between them. According to the National Board of Boiler and Pressure Vessel Inspectors, the following are the definitions of each.

- *Safety Valve – This device is typically used for steam or vapor service. It operates automatically with a full-opening pop action and recloses when the pressure drops to a value consistent with the blowdown requirements prescribed by the applicable governing code or standard.*
- *Relief Valve – This device is used for liquid service. It operates automatically by opening farther as the pressure increases beyond the initial opening pressure and recloses when the pressure drops below the opening pressure.*
- *Safety Relief Valve – This device includes the operating characteristics of both a safety valve and a relief valve and may be used in either application.*
- *Temperature and Pressure Safety Relief Valve – This device is typically used on potable water heaters. In addition to its pressure-relief function, it also includes a temperature-sensing element which causes the device to open at a predetermined temperature regardless of pressure. The set temperature on these devices is usually 210°F.*

The following is a list of common mistakes with boiler safety relief valve installations.

Relief Valve Piping - The boiler contractor installed a 1 ½" x 3/4" bushing on the outlet of the safety relief valve. Instead of 1 ½" pipe, the installer used 3/4" pipe to the floor drain. When asked about it, he answered that he did not have any 1 ½" pipe but had plenty of 3/4" pipe. I explained and then had to show the disbelieving contractor the code that states that the relief valve discharge piping has to be the same diameter as the relief valve outlet. See 2012 International Mechanical Code, 1006.6. By reducing the discharge pipe size, the relieving capacity of the safety valve may not be adequate to properly relieve the pressure inside the boiler, causing a dangerous situation. The code also states that the discharge material shall be of rigid pipe that is approved for the temperature of the system. The inlet pipe size shall be full diameter of the pipe inlet for the relief valve. Some manufacturers suggest using black iron pipe rather than copper tubing. If using copper, it should have an air space that allows expansion should the relief valve open to avoid the accident that I referenced above. The discharge piping has to be supported and the weight of the piping should not be on the safety relief valve. Valves are not permitted in the inlet piping to or discharge piping from the relief valve. If you are using copper tubing on discharge piping, verify that there is room for expansion.

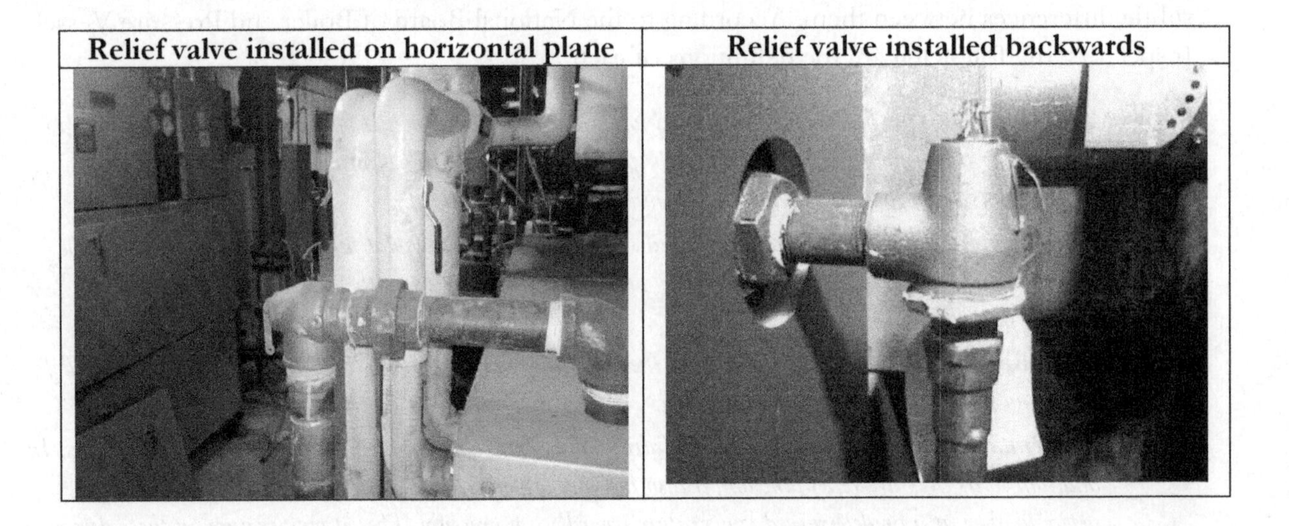

Relief valve installed on horizontal plane	Relief valve installed backwards

Installation – Read the manufacturer's installation manual as each may have different requirements. For instance, Conbraco requires that the discharge piping must terminate with a plain end and use a material that can handle temperatures of 375^{0}F or greater. This will preclude PVC or CPVC pipe for the discharge piping. The instruction manual for their model 12-14 steam relief valve stipulates that you cannot use a pipe wrench to install it. I am not sure what they expect the installer to use but that stipulation would be good to know.

I once visited Boiler Utopia as the floor was clean and waxed. All the pipes were covered and exposed pipes were painted. There were large stickers detailing what was inside each pipe as well as directional arrows. Nothing was stacked next to the boilers. Yellow caution lines were painted on the floor around each boiler. I was in heaven. As I walked around the rear of the boiler, something clicked and triggered a warning bell. The discharge of the relief valve piping was about 6" from the floor but instead of a plain or angled cut end, the pipe had a threaded pipe cap on the termination. I asked the maintenance person about it and he said that the valve was leaking all over his newly waxed floor and this was the only way he could stop it. When I said that the discharge pipe should not have been threaded, he explained that it was not threaded and had to take it to the local hardware store to have them thread it. I informed him that the cap had to be removed. We cut the pipe on an angle to prevent this.

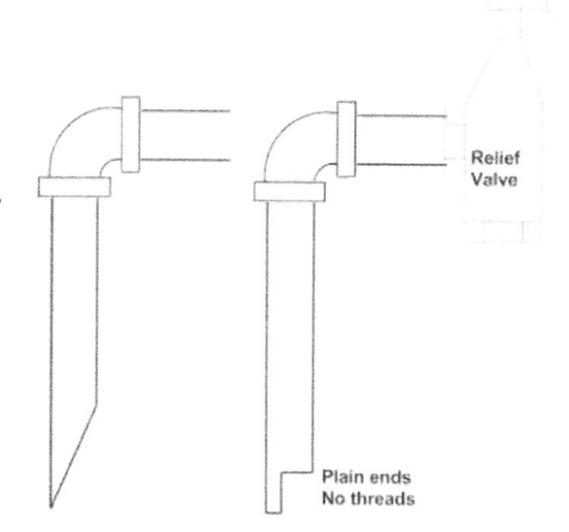

Steam Boiler Most manufacturers recommend a drip pan ell on the discharge of the steam boiler relief valve to eliminate the weight of the discharge piping on the relief valve. Some codes require the discharge to be vented outdoors. If you are going to manifold the discharge piping, the discharge piping should be sized so the cross sectional area is equal to the cross sectional are of all devices connected to the pipe.

Testing "How often do you test the relief valves?" I will ask the attendees in my classes. Most do not make eye contact and when I follow up with "Why are they not tested?" I often hear that opening the relief valve will cause it to leak. I suggest that you refer to each manufacturer's directions for testing. For instance, one will recommend once a year while another recommends twice a year. One manufacturer says that "Safety/relief valves should be operated only often enough to assure that they are in good working order." I am not sure what that even means. You want to also verify the proper test procedure as some manufacturers will want the relief valve tested when the boiler is at 75% of the rated pressure or higher of the relief valve. For example, you should test the steam boiler relief valve when the boiler pressure is at 11 psi or higher. If your system is set lower than that, you will need to increase the boiler pressure and then lower it when the test is complete.

Relief Valve Leaking A leaking relief valve could be a future service call. The cause of the leak should be investigated. On a hydronic boiler, a weeping relief valve could indicate the following:

- Plugged piping to the expansion tank
- Flooded expansion tank
- Defective water feeder
- Excessive boiler pressure
- Undersized expansion tanks
- Defective bladder in expansion tank

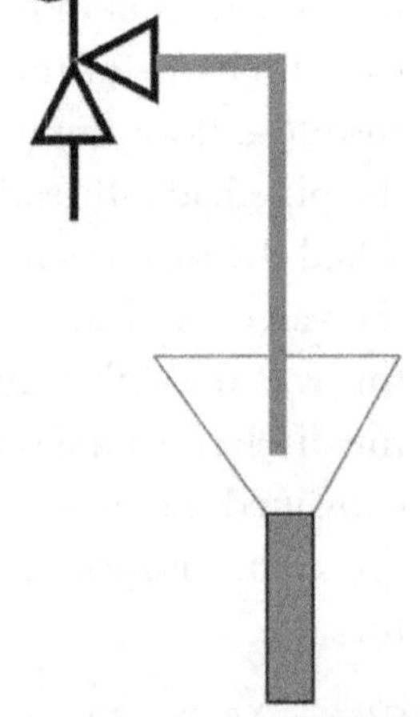

Relief Valve Sizing The sizing of a boiler relief valve is a complex calculation that assures the boiler will not be allowed to have more than 10% over pressurization when the relief valve is fully open. It factors in the discharge area of the relief valve, coefficient of the substance inside the boiler, and much more. It is much easier to ask the boiler manufacturer what size relief valve they require. The relief valve must be set lower than the Maximum Allowable Working Pressure (MAWP) of the boiler. This should be stamped on the boiler. The low pressure steam boiler relief should be set for 15 psi. The relieving capacity should be more than the output of the boiler. In many cases, the relieving capacity is greater than the input of the boiler.

How to test your courage. We sold a high pressure steam boiler to a federal government facility. Part of the startup process was to "jumper" the operating and limit controls to see if relief valve would relieve the pressure inside the boiler. As the steam pressure rose, so did my blood pressure. When the relief valve finally opened, I jumped and gasped loudly. It was awesome. The noise sounded like a jet engine roaring. The steam plume was nearly invisible on the outlet of the relief valve piping for about 20 feet.

Be kind to the safety relief valves as they may save a life some day and it could be yours.

Relief Valve piped solid to ground	Relief Valve piped solid to ground

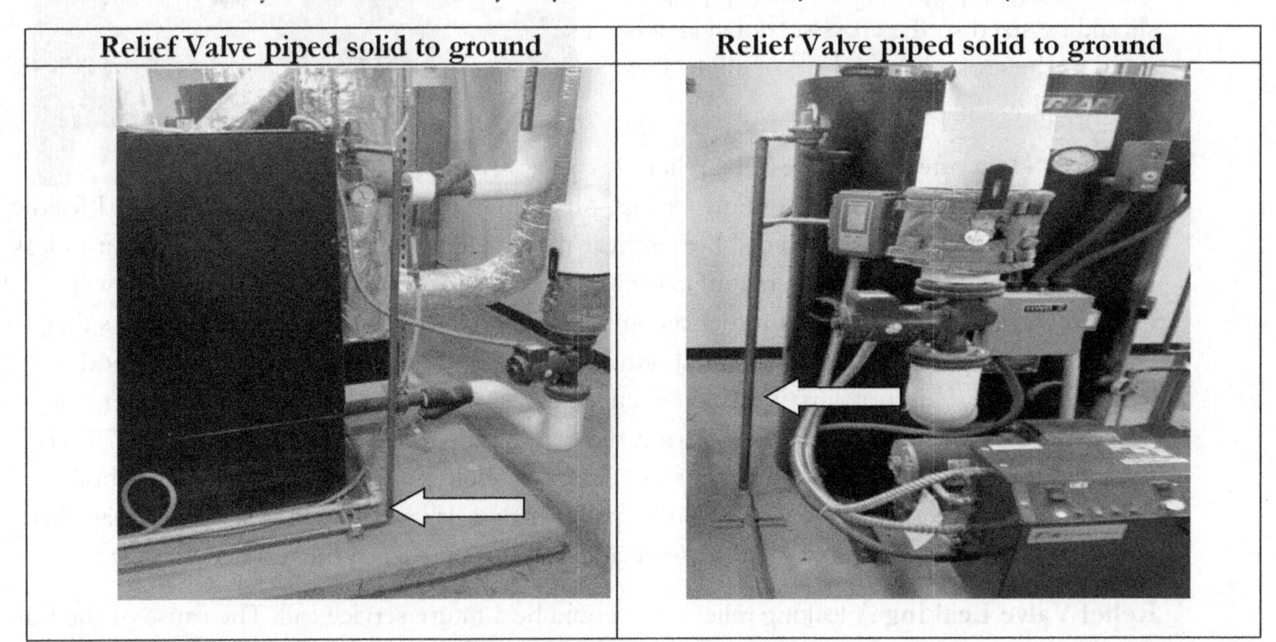

Flue & Breeching

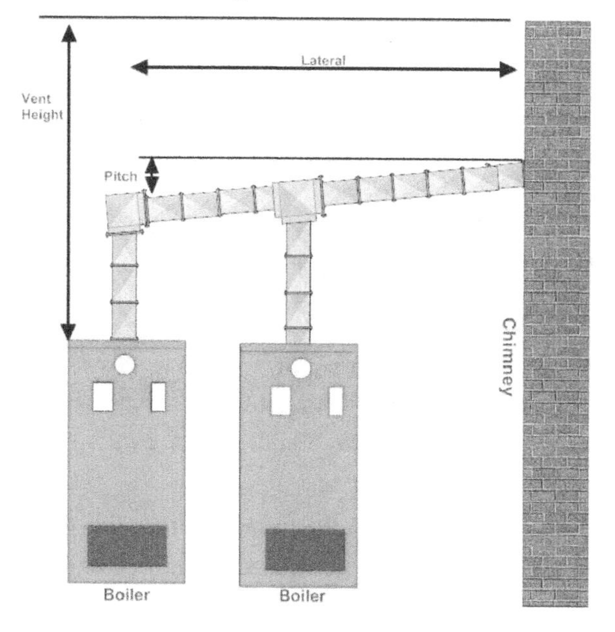

Verify the integrity of the flue and that it is installed as per the manufacturer's recommendations. Check all joints and note the size and pitch of the flue. Verify that the stack is the correct category for the boiler. I heard of a Category 1 boiler vented with Category 3 stack. The stack was not designed to be operated in a negative condition and collapsed shutting off the flow of flue gases from the boiler. See if there is evidence of corrosion in the flue or chimney. This could be caused by either the boiler operating at temperatures that are too low or could even mean that there is a leak in the boiler. Verify that the joints are connected properly. If the flue is Snaptite with crimped ends, the joints should be connected and held together with screws. If the boiler is either a Category III or IV, verify that the flue is of the correct type and that is its air tight as it is a positive flue and could leak flue gases.

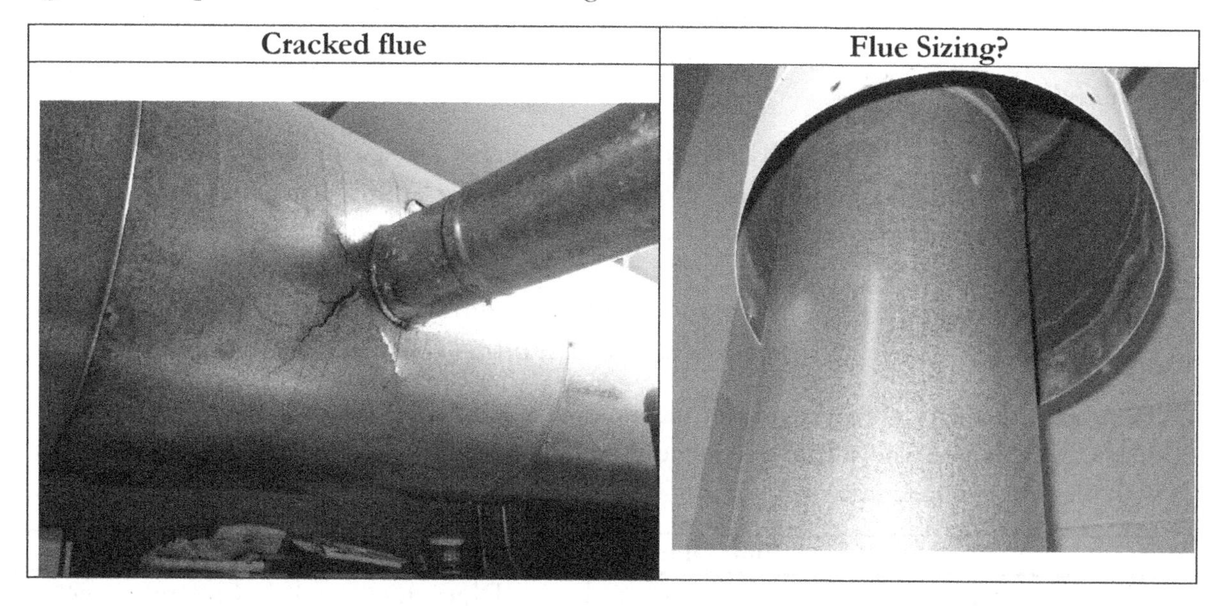

Cracked flue | **Flue Sizing?**

Flue gas spillage	Installer used 13 elbows to vent water heater. Courtesy: Bacharach

Flue Vent Problems Please notice the discoloration on the boiler as this indicates a dangerous condition. It is a safe bet that carbon monoxide is being vented into the boiler room so use extreme caution. It shows that the flue gases were not venting through the boiler. Rather, they were rolling out the front of the boiler into the boiler room. Check the draft diverter on the water heater for venting problems as indicated by rust on the top of the water heater. White flakes inside the flue will indicate either a leak or the flue gases condensed. The white powder is what is left after the moisture evaporates. The cause should be investigated. It may be a leaking air-handling unit in the room, negative conditions in the space, plugged flue passages or inadequate combustion air.

Flue Diameter Some service calls will require you to verify the size of the flue. Measuring the size of the existing flue is sometimes a challenge, especially a large one. You find yourself squinting and trying to "eyeball" each starting point. An easier way to find the diameter of the flue is to measure the circumference of it, using a wire. You then straighten the wire and measure the length. The following table is the circumference for different size round flues. To find the circumference of a circle, the formula is ***Circumference = Diameter x π***

To find the diameter, simply divide the circumference by π or 3.1416. When sizing B Vent, the inside diameter is usually one to two inches less, depending on wall thickness. To be sure, check with the vent manufacturer.

Round Flue Diameters in Inches Single Wall			
Diameter	Circumference	**Diameter**	Circumference
12	37.70	26	81.68
14	43.98	**28**	87.96
16	50.27	**30**	94.25
18	56.55	**32**	100.53
20	62.83	**34**	106.81
22	69.12	**36**	113.10
24	75.40	**38**	119.38

7 Times Rule A common mistake installers make is when they remove the boiler flue and leave the water heater flue in the old chimney. This is referred to as an orphan. There is a rule of thumb when checking the flues called the "7 Times Rule" which is: The flow area of the largest common vent or stack shall not exceed seven times the area of the smallest draft hood outlet. Since most water heaters use a 3" flue, the largest area to connect the water heater should be 49" in area or a 7" round one. The following is a chart that shows the largest common round flue that each can be connected to.

7 Times Rule Round Flue			
Smallest Draft Hood Outlet	**Largest Common Flue**	**Smallest Draft Hood Outlet**	**Largest Common Flue**
3"	7"	**9"**	22"
4"	10"	**10"**	26"
5"	12"	**12"**	30"
6"	14"	**14"**	36"
7"	18"	**16"**	42"
8'	22"		

If you are connecting to a square or rectangular chimney, you will need to estimate the areas. The following is a chart that shows the sizes that the vent can be connected to:

7 Times Rule Rectangular Flue		
Smallest Draft Hood Outlet	**Area of Vent**	**Largest Common Vent Area**
3"	7.06"	49"
4"	12.56"	88"
5"	19.64"	137"
6"	28.27"	198"
7"	38.48"	269"
8"	50.27"	352"
10"	78.54"	550"
12"	113.1"	792"
14"	153.94	1,078"
16"	201.06"	1,407"

Allow me to show you an example: If you remove the boilers from a chimney that is 12" x 12" square and want to see if we can leave an old water heater with a 4" flue outlet. The 12" x 12" chimney is 144" This is greater than the rule of 7 sizing for the 4" flue which is 87.92". On this job, we would have to make provisions for the water heater flue. A chimney liner would most likely have to be installed here.

Horizontal vs. Vertical Another rule to remember is that the horizontal vent must be no more than 75% of the vertical height of the flue. If using B Vent, the horizontal length can be the same length as the height of the chimney, as per International Fuel Gas Code, 2006 503.10.9.

Condensing Boiler Flues Most condensing boilers require a Category IV positive flue. This flue is a positive flue with condensing boilers and since it is positive, they cannot be combined with other flues like the older boilers did. Each boiler will require a separate flue, which increases the installation costs. In addition, the side wall of the building looks like an old pirate ship that had cannons every few feet if there are multiple high efficiency appliances.

When venting these boilers, most cannot use a common breeching. Since each boiler operates with a positive pressure in the flue, this could force the flue gases out of the idle boiler and damage the boiler. In addition, the flue gases could flow out of the idle boilers and into the boiler room. I have seen them combined when used with a chimney top exhaust fan to make sure the flue is negative.

"My bushes are ruined", the wife of the CEO of a large medical facility said angrily to her husband. The cause of her angst was that the new condensing boilers were vented directly above the bushes she had planted. When the flue gases condense, acids are formed when mixed with the water vapor. The flue gases, containing acids, landed on the bushes that the CEO's wife had purchased and planted. The bushes were destroyed and the CEO was furious as you can imagine. They wanted to get rid of the boilers and replace them with standard boilers that vented through the chimney. Be careful where you vent the condensing boilers. This cost the Director of Maintenance his job.

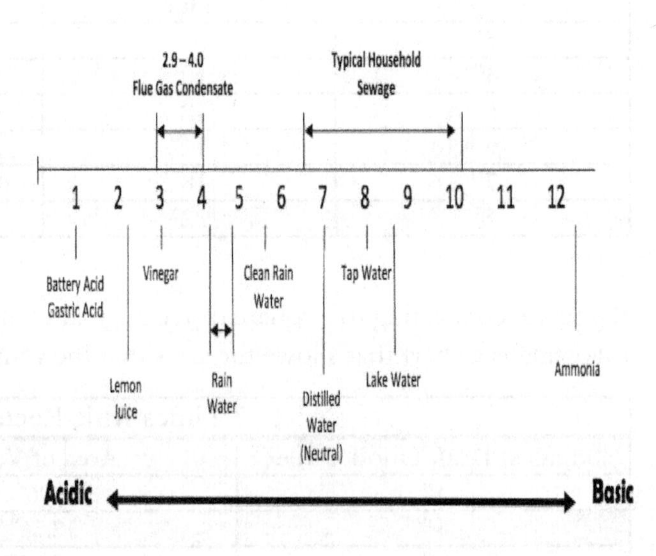

Credit: Mike Bernasconi, Neutrasafe Corp.

Flue gas condensation When a boiler condenses, it means that the flue gas temperature drops low enough to fall below the dew point temperature. When this happens, the moisture that was part of the flue gas falls out and accumulates inside the boiler. A condensing boiler will produce 2 cubic feet of water for each cubic foot of gas burned. The flue gases also have levels of sulfur and carbon dioxide, creating sulfuric and hydrochloric acids when they combine with the water vapor. Due to the acid formation, the pH lowers. It is typically between 2.9 to 4.0 A reading of 7 is neutral. While it appears to be an insignificant amount, pH readings are logarithmic, which means that a reading of 6 is ten times more acidic than a reading of 7. A reading of 5 is 100 times more acidic than a reading of 7. So you can see that a reading is 2.9-4.0 is vastly more acidic than the neutral reading of 7. Many ask how that would affect the piping. Consider this, when the pH level is greater than 8 or more which is alkaline, copper pipes form a copper oxide film, protecting them from corrosion. When the pH is lower than 8, the copper oxide film in the copper pipe dissolves and the pipe is vulnerable to corrosion, resulting in leaks.

Flue Information

Vent Categories				
	I	II	III	IV
Vent Pressure	Negative	Negative	Positive	Positive
Temperature	>275^0	<275^0	>275^0	<275^0
Efficiency	<84%	>84%	<84%	>84%
Gas tight	No	No	Yes	Yes

The Products of Combustion Produced When One Cubic Foot of Gas is Burned	
One Cubic Foot of Gas Burned Produces	8 Cubic feet of nitrogen
	2 Cubic feet of water vapor
	1 Cubic Foot of Nitrogen

Typical Vent Temperature Ranges		
Venting Material	Temperature Ratings	Fuel
AL29-4C Stainless	0 - 480^0 F	Gas
B and BW Vent	0 - 550^0 F	Gas
L Vent	0 - 1,000^0 F	Oil
Factory Built Chimney	500^0 - 2,200^0 F	Oil/Gas
Masonry Chimney	360^0 - 1,800^0 F	Oil/Gas
Verify with manufacturer		

Condensing & Ignition Temperature of Various Fuels

Fuel	Condensing Temperature	Ignition Temperature
Natural Gas	250 ^{0}F	1,163 ^{0}F
#2 Oil	275 ^{0}F	600 ^{0}F
#6 Oil	300 ^{0}F	765 ^{0}F
Coal	325 ^{0}F	850 ^{0}F
Wood	400 ^{0}F	540-1,100 ^{0}F

Minimum Flue Gas Temperatures for Category 1 Boilers

Fuel	Minimum Flue Temperature for Non Condensing Appliances
Natural Gas	265°F plus 1/2°F for each foot of stack or breeching, including horizontal and vertical runs
#2 Fuel Oil	240°F plus 1/2°F for each foot of stack or breeching, including horizontal and vertical runs

Acid Rain and Stack Temperature

Fuel	Acid Dew Point Temperature	Minimum Allowable Stack Temperature for Non Condensing Appliances
Natural Gas	150	250
#2 Fuel Oil	180	275

Boiler Draft Draft is the measurement of the velocity of the flue gases from the boiler to the chimney and is typically measured in inches of water column or "W.C." The higher the draft, the faster the flue gases are travelling through the boiler. If the velocity is too slow, you could have excessive heat inside the boiler and have spillage of the flue gases into the boiler room. Inadequate draft could also damage the burner. If the velocity is too high, the flue gases cannot give up the heat into the boiler lowering the efficiency. It could also distort the flame pattern, causing Carbon Monoxide CO to form. The draft should be measured during the service call. I like to measure it when the stack is cool to see how much spillage occurs and also when the stack is hot to see if it is excessive. The spillage should only last a few minutes until the stack is warm.

Rule of thumb. For every 0.01" w.c., the excess draft can be reduced, the fuel consumption is reduced by 1%

Commonly Accepted Draft Measurements

Type of Heating System	Overfire Draft	Stack Draft
Gas, Atmospheric	Not Applicable	-.02 to -.04" WC
Gas, Power Burner	-.02" WC	-.02 to -.04" WC
Oil, Conventional	-.02" WC	-.04 to -.06" WC
Oil, Flame Retention	-.02" WC	-.04 to -.06" WC
Positive Overfire Oil & Gas	+.4 to +.6	-.02 to -.04" WC

Draft Controls If the building stack or chimney is over 30 feet tall, a draft control is recommended when using Category 1 appliances. A draft control could be anything from a draft diverter to a barometric damper up to a sequencing draft control. The barometric dampers are more common due to their lower installed cost. When using a barometric damper, the red stops on a barometric damper should be removed if the boiler is firing with only natural gas. The red stops remain if firing with fuel oil. The stops will not allow the barometric damper to open and spill flue gases back to the room if fired with oil but the flue gases on a gas burner can be vented into the boiler in the event of a chimney blockage. It amazes me that this is permitted. That is why I recommend installing a spill switch on the barometric damper. This will shut off the burner if flue gas spillage is detected, making it a much safer boiler room. Spill switches should be used anytime there is a barometric damper or a draft diverter. Many of the new boilers and furnaces have integral spill switches.

Sequencing draft controls are typically installed in larger facilities, such as hospitals or industrial facilities. They are more accurate and much more expensive than a barometric damper. A sequencing draft control consists of an electrically operated damper on the outlet of the boiler, a draft-sensing device and a draft control. The draft control will modulate the draft damper to maintain the desired draft set point. They do require more service more than a barometric damper.

Induced Draft Fan Some commercial boilers use an induced draft fan to pull the products of combustion from the boiler. This should be checked during your service call. I once had a fan that was completely destroyed because the boiler water temperature was too low. It was a very expensive repair. The one shown in the picture is a belt driven model.

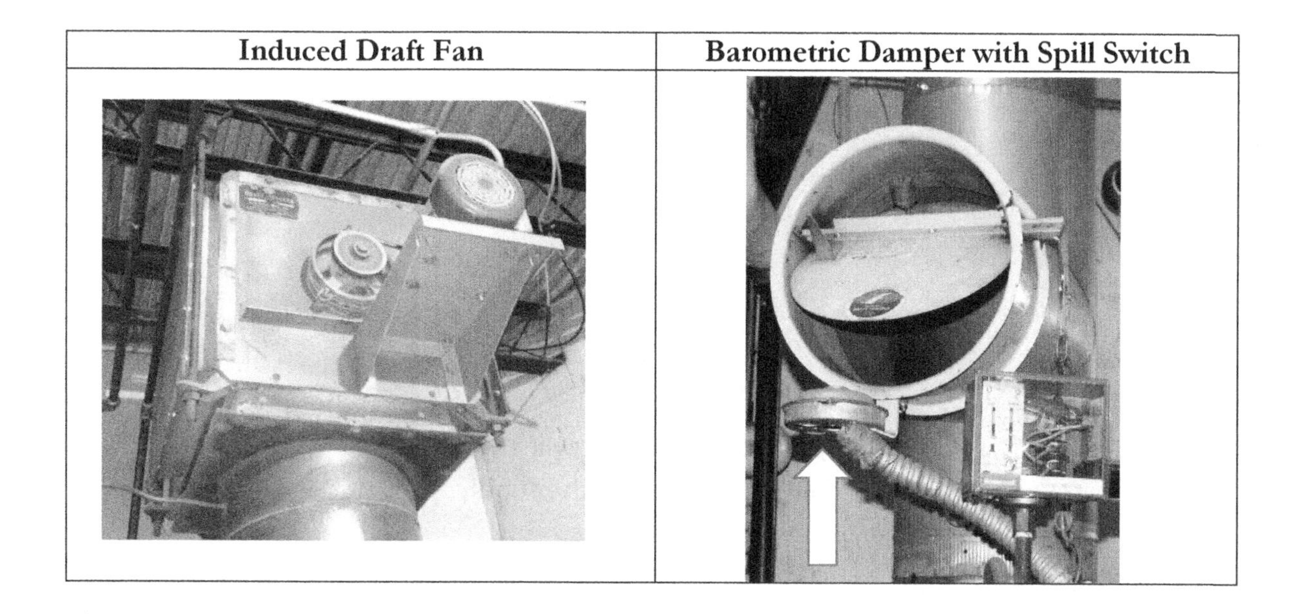

Induced Draft Fan | **Barometric Damper with Spill Switch**

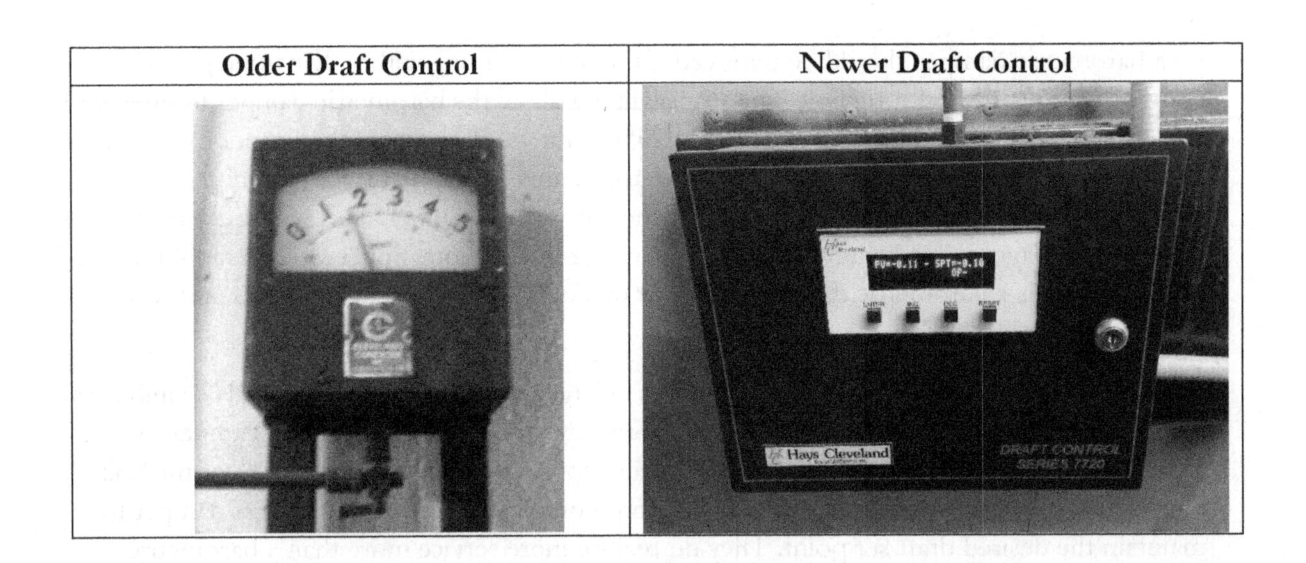
Older Draft Control
Newer Draft Control
0 1 2 3 4 5
Hays Cleveland
DRAFT CONTROL
SERIES 7720

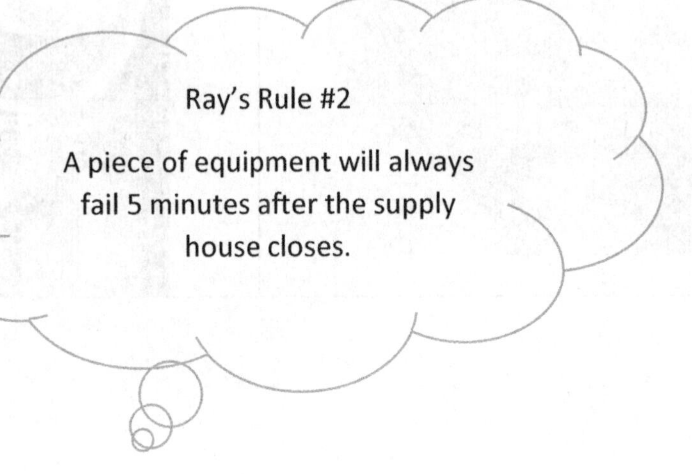
Ray's Rule #2
A piece of equipment will always fail 5 minutes after the supply house closes.

Some Random Thoughts on Boiler Control

I will discuss in further detail boiler control in the subsequent sections on hydronic and low pressure steam.

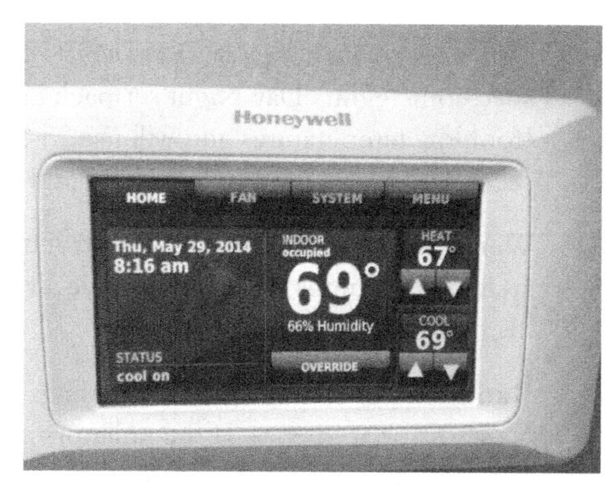

Boiler Sequencing What is the best way to sequence boilers? I was discussing the boiler sequencing on a new project with the "Out of town" control expert that the client had hired.

"I always change the lead boiler each time there is a call for heat. In that way, we even the run time of all the boilers." He said with a condescending smile.

I respectfully disagreed with his logic. I explained that with standard efficiency boilers piped in a primary secondary arrangement, you should consider firing the same lead boiler for an extended period of time. When you fire a different boiler each time, you are firing into a cold boiler since each boiler is isolated. This leads to flue gas condensation and erosion of the fire side of the boilers. If you are firing into a boiler that is already warm, the flue gases will be less likely to condense. This will extend the life of the system. I prefer to have one boiler as the lead boiler for a month or two or even a full heating season. After that, I would change the lead boiler to the next boiler and cycle the previous lead boiler as the last boiler to fire.

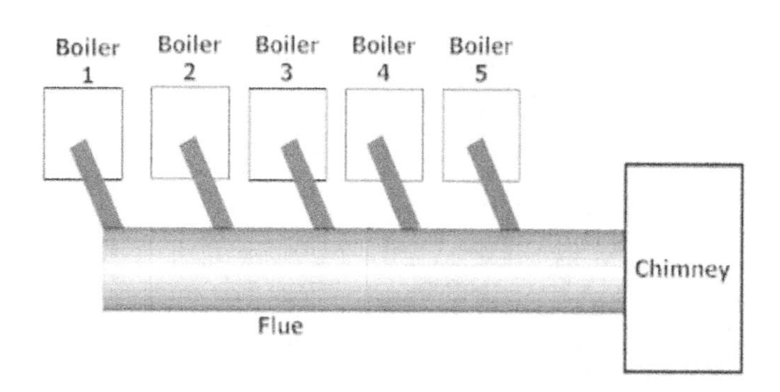

Another consideration of boiler sequencing is the boiler arrangement of non-condensing boilers and their proximity to the chimney. In the picture to the right, you will see that this project has five boilers. If I were controlling this project, I would have Boiler #5 as my lead boiler. This will preheat the flue and chimney. If Boiler #1 was the lead boiler under light load conditions, we could have the flue gases condense before they reach the chimney.

Outdoor Heating Design Temperature - This is the design temperature that engineers use when sizing a heating system. It used to be the coldest temperature that the locale experienced. It is now the temperature that occurs 2 1/2 % of the time. This means that in a typical winter, the outside temperature will be at or cooler than the design temperature 2 1/2% of the time or about 100 hours per year.

Thermostat The easiest way to control a boiler is to connect it to a thermostat. While this may not be very sophisticated, it works on small buildings.

When using a night setback control, the recovery time is longer for radiators. In severe weather, you may have to start the occupied cycle several hours before the actual occupied time to allow the radiators to heat the space. On bitter cold days, consider eliminating the night setback all together. When using a thermostat, there is no outdoor reset of the water temperature.

Should we even use a night setback control? I found this interesting quote by York Shipley that may shed some light. "Day Night set-back operating systems amplify the potential for boiler damage from low temperatures and will use 25-50% additional fuel than a conventional primary secondary loop system working around the clock."

Warm Weather Shutoff This control uses an outdoor air sensor that will shut off the heating system and pumps when the outside air temperature reaches a certain degree, typically 55-60 degrees F.

Radiator Covers Many older buildings or homes enclosed the radiators for either safety concerns or aesthetics. If the radiators are enclosed, the heat output is affected by as much as 30%. The chart below shows the effect of radiator enclosures on the heating capacity of the radiator. *Is the radiator really a radiator?* Did you know that the heat output of a radiator is 60% convection and only 40% radiation?

Radiator Enclosure Effects	Radiator Enclosure

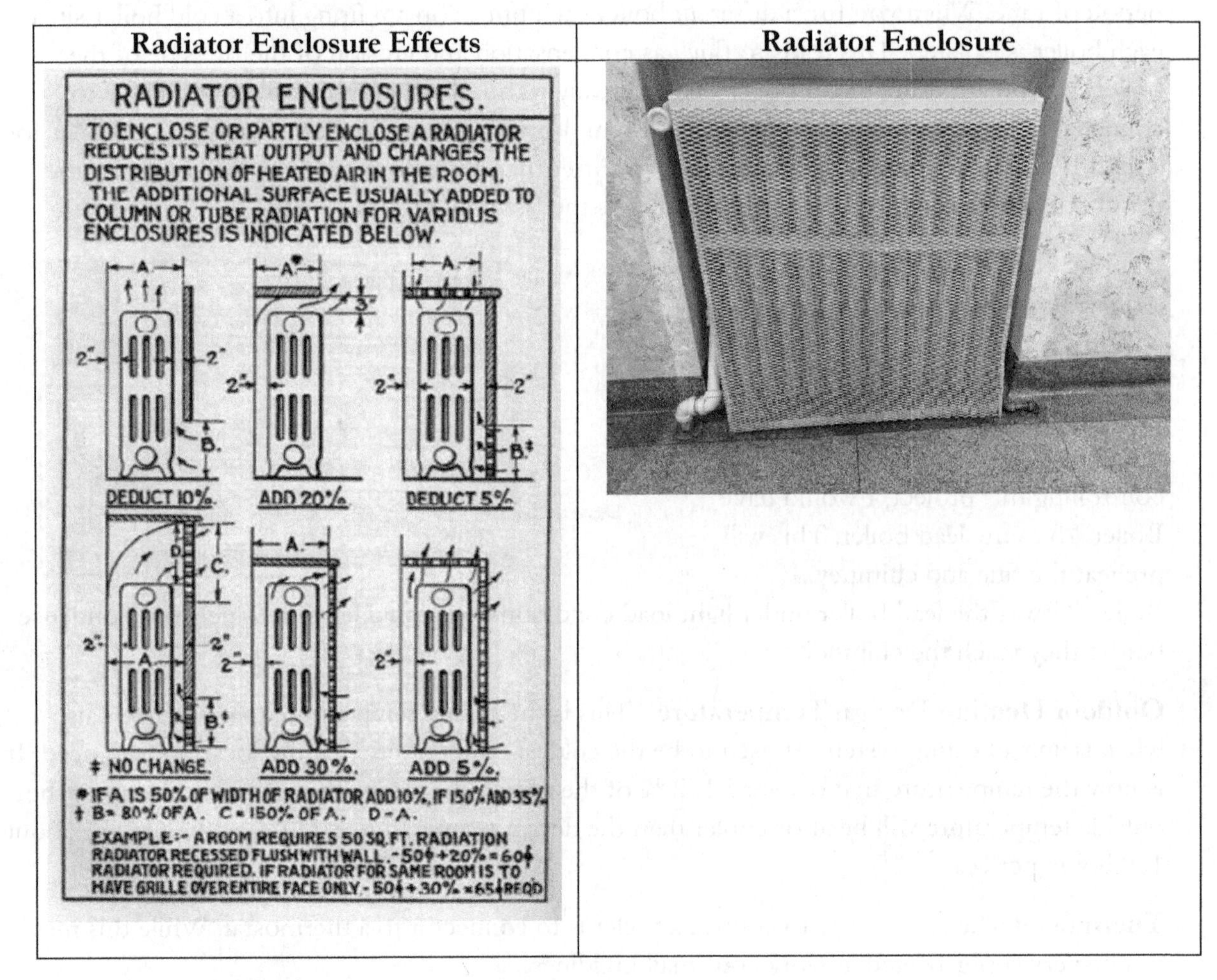

Combustion Analysis

Understanding Boiler Efficiency Boiler efficiency is sometimes confusing as there are several similar sounding terms that have very different meanings.

Combustion Efficiency This is the efficiency of the burner while it is operating and is measured with a combustion analyzer. When the analyzer shows that a boiler is 80%, that means that 80% of the fuel burned is being transferred to the boiler and 20% is being sent up the chimney.

Boiler Efficiency This is the measurement of the boiler's ability to transfer heat to the system. This may be influenced by the connected piping, scale buildup on the water side, boiler sizing and cycling, and jacket loss.

Seasonal Efficiency This is the efficiency of the heating system over an entire heating season. System sizing, boiler cycling and heating transfer ability of the system factors into this calculation.

Understanding Combustion What does the combustion efficiency reading mean? When servicing burners, you become familiar with readings from a combustion analyzer. The analyzer uses a sensing tube in the flue outlet of the boiler to measure the efficiency of the burner as well as the emission gases.

History of Combustion Testing The early type of analyzers used a chemical process to detect the levels of carbon dioxide or oxygen in the flue gases using the "Orsat" method of volumetric analysis. The unit would have either potassium hydroxide for testing for carbon dioxide or chromous chloride for testing for oxygen. These units were called "dumb bells" in the field. The chemicals inside the tester would absorb the sample gas and show a reading to the user. The disadvantage to this type of tester is the accuracy and speed. If you performed three tests, you got three different results. In addition, the tests took forever to complete. It was still better than simply trying to eyeball the flame.

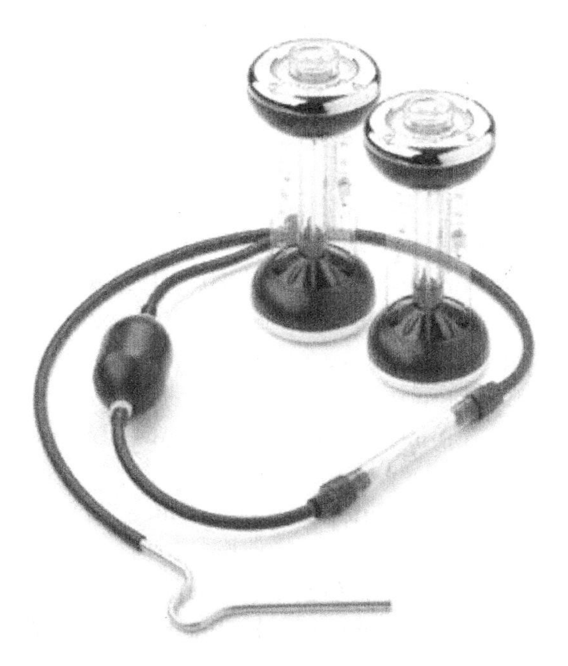

Bacharach Fyrite Gas Analyzer

The next analyzers used to test for carbon dioxide only in the flue gases because early oxygen sensors had an extremely short shelf life. The disadvantage to testing only for carbon dioxide was that you never knew if carbon monoxide was present. You could have great CO2 readings but the boiler could be spewing lots of dangerous carbon monoxide or CO out the stack.

The new analyzers are much easier to use and will give a continual readout which will make adjusting the fuel to air easier. In addition, you could get readings of the O2 and CO levels in the flue gases, making it a safer piece of equipment.

Combustion Analyzer Care The combustion analyzer is a very expensive piece of equipment and should be used with care. Many of the units do not react well to cold weather and should be kept inside rather than left in the cold service vehicle where it could be damaged. When you are done with your testing, make sure all the water from the flue gases is drained from the analyzer trap as it contains acids that could prematurely age the unit. The unit should be calibrated regularly. Some contractors prefer to perform the calibrations themselves but I like sending them back to the manufacturer so I can have it certified to avoid any legal issues if a problem arose on a job.

Combustion analysis Let us pretend that we have invented the perfect burner. It is 100% efficient. To get perfect (100%) efficiency, our burner needs 10 parts of air to 1 part of gas. The only problem is that our perfect burner will require constant attention. If the ratio of fuel to air changes slightly, we could have problems. There is no room for errors. What happens if the blower wheel accumulates some dirt, the linkages slip or even if the boiler room temperature changes by 20^{0}F? If the amount of air supplied to the burner drops, we could soot our boiler, which will be dangerous and increase our fuel costs. (In reality, our burner does not require air but actually requires oxygen. To introduce oxygen into our burner, we need to bring in air since it contains oxygen. *Did you know that for every cubic foot of oxygen that the burner requires, we need to bring in 4.78 cubic feet of air?*

Since we cannot be there 24 hours a day to monitor our burner fuel to air ratio, we decide to add a little extra air as a safety factor. After all, we need to go the bathroom or eat at some time. We increase the amount of air from 10 to 13 parts of air. Our new air to fuel ratio is now 13 parts of air to every part of gas and we have 30% extra or "Excess Air". Our burner efficiency is sacrificed slightly but we are able to have a system that heats without having to be monitored all the time. Even if the blower wheel gets dirty or the linkages slip slightly, we should have a safe boiler. On commercial boilers with power burners, they are designed to supply between 12 and 15 parts of air for each part of gas. In some industrial facilities, they can reduce the fuel to air ratio even lower than 20% by using an oxygen or O2 trim system. This system uses a sensor in the flue to monitor oxygen content while the burner is running and adjusts the air damper.

Atmospheric boilers use a slightly higher amount of air. The primary and secondary air for an atmospheric boiler is through the base. The draft diverter is used for dilution air or tertiary air. This dilution air is fed into the draft diverter so that it does not steal the heat from the boiler.

A combustion analysis should be performed a minimum of once per year. It does not matter when it is done if it is not done incorrectly. The picture to the right shows the hole for the combustion analyzer probe was installed above the draft diverter. When we tested the boiler, the air reading above the draft diverter was normal but below it was very low and the boiler was sooting.

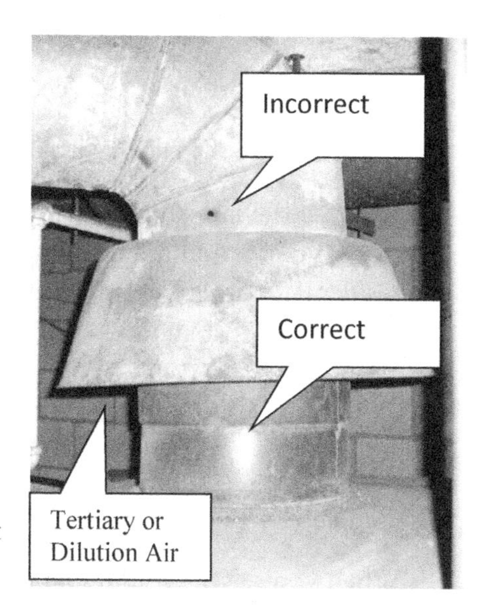

When is the best time to perform the combustion analysis: Fall, Winter or Spring? To get the proper readings, the boiler should run for 15 minutes prior to any adjustments. This allows the flame to stabilize. When performing the combustion analysis in the Fall and Spring, the boiler may shut off before any adjustments could be completed. The best time to perform the combustion analysis is when the building is under a heavy load, such as mid winter.

A combustion analysis adjusts the optimum amount of air for your boiler and burner. Too much air and you are wasting fuel. Too little and you could soot the boiler. A combustion analysis should also be performed if the following items are done: clean or replace blower wheel or motor, replace gas valve or gas pressure regulator or the linkages slipped.

Fuel oil requires 1,400 cubic feet of air for each gallon of fuel, rated at 140,000 Btuh, for perfect efficiency.

Oxygen or O2 The air we breathe and use for burner combustion contains about 78% nitrogen, 21% oxygen, and about 1% argon and trace elements of other gases, including moisture. When we adjust the air to a burner, we try to adjust the least amount of air that we can for best efficiency. This is called the Air-Fuel ratio or AFR. If we were able to have perfect efficiency, it would be called Stochiometric. To get that perfect efficiency, we would have about 10 parts of air for one part of natural gas. That is not realistic because the moisture content in the combustion air or the Btu content in the gas could change, affecting the burner efficiency. If the burner has too much air or insufficient fuel, that is called Lean and lowers the efficiency. If the burner has too little air or too much gas, it is called Rich and that could lead to sooting and high emissions of the burner, including carbon monoxide. To make sure we have a safe flame, we add some extra or Excess Air. On industrial boilers with sophisticated monitoring, we may get away with about 15% excess air. That may look like this:

1 cubic foot of gas

10 cubic feet of air for perfect efficiency

1 ½ cubic feet of excess air or 15% excess air

Many of the industrial boilers use an oxygen or O_2 trim system that constantly monitors the oxygen content in the flue gases and adjusts the Fuel-Air Ratio. This is rare on the commercial or residential burners, so we usually try for more air to make it a safer flame. The commercial boiler is usually set for about 20% excess air at high fire. It would look like this:

1 cubic foot of gas

10 cubic feet of air for perfect efficiency

2 cubic feet of excess air or 20% excess air

When you see the reading on the combustion analyzer, it will read Oxygen percentage. That reading is amount of oxygen as percent of the flue gas volume. For instance, a reading of 4% O_2 means that 4% of the volume of flue gases is oxygen. A typical reading would be 2-6%. On a commercial burner, I like getting a reading of about 4%. I think that this gives me a little bit of a safety factor in case the blower wheel gets dirty. If the wheel gets dirty, you will provide less air and the mixture could be too rich causing elevated carbon monoxide levels and the chance of sooting the boiler is increased.

The nitrogen that is part of the air simply goes through the combustion process. It does not affect the fuel to air ratio. If anything it reduces the temperature of the flue gases.

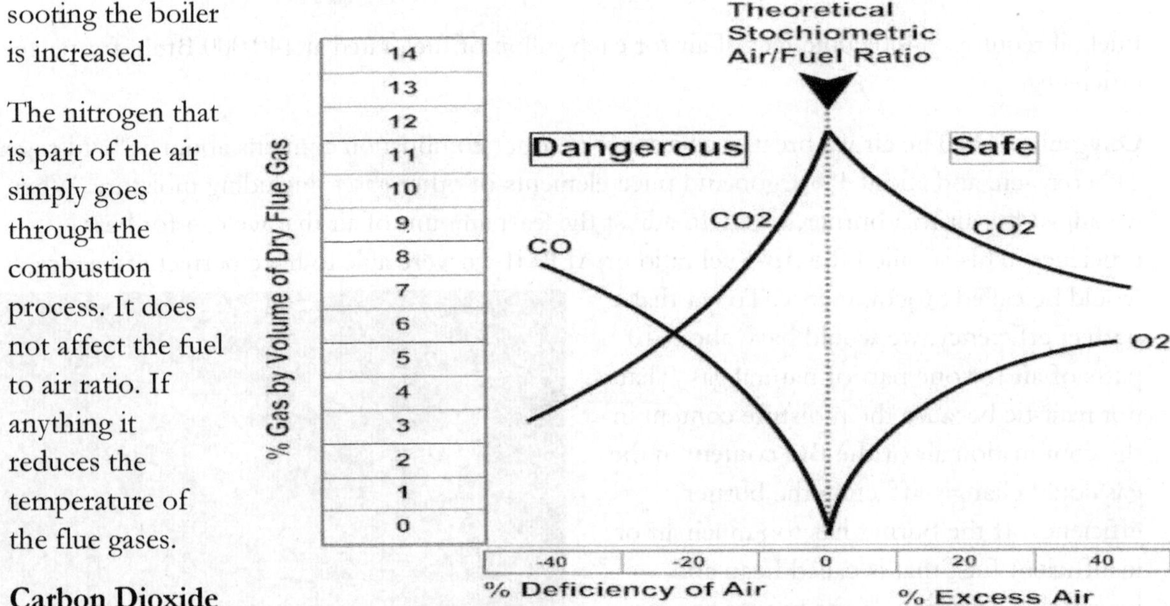

Carbon Dioxide or CO_2 is also a byproduct of combustion. The early combustion analyzers used to measure the carbon dioxide only levels in flue gases to determine the efficiency of the burner. The drawback to measuring only carbon dioxide or CO_2 was that Carbon Monoxide could be present in the flue gases. If you look at the theoretical chart above, you will see that oxygen is on the right side of the chart while carbon monoxide or CO is on the left. If you are measuring only CO_2, you cannot determine which side of the chart you are on. You could have a good efficiency with acceptable CO_2 levels and spewing carbon monoxide up the flue. Many burner manufacturers look for a reading of between 8 ½ to10% in the flue gases.

Carbon Monoxide or CO is a dangerous byproduct of combustion that is generated by incomplete combustion. It is measured in parts per million or PPM rather than a percent of flue gas. To see how tiny a part per million is consider this. One part per million is:

- One day in 2,739 years
- 1 inch in 16 miles
- 1 drop of vermouth in 80 fifths of gin

Carbon Monoxide is like a crotchety old man that is set in his ways and hates change. Each time the burner flame changes position, the CO level will rise then should drop back to its settings, almost like a protest for being moved. It is normal but make sure it does not rise too far. If it does, you may have to adjust the linkage stroke. Carbon monoxide is also produced if the flame pattern is disturbed, such as when the draft is incorrect or there is rollout of the flue gases.

When adjusting the fuel to air ratio, I like to get the carbon monoxide levels as close to 0 as I can. In many instances, we cannot so I try getting it below 100 parts per million air free. The limit in flue gases is 400 CO Air Free according to ANSI Z 21.1. If you measure just the CO without factoring in the air free portion, this is referred to as "as measured" carbon monoxide.

CO Air Free When measuring CO in the flue gases, we use a term called CO Air Free. What that means is that you subtract the excess air from the calculation to measure the actual CO amount in the flue gases. If you have an elevated carbon monoxide level in the flue gases, some techs are tempted to add more air to try lowering the CO. This does not really work and the CO Air Free takes away the excess air to get the actual reading. For example, let us say that you and a friend went to Starbucks and each purchased a Grande coffee which is about 16 ounces. You take your coffee black but your friend adds cream and sugar. You would both have 16 ounces of coffee but your friend's coffee would be diluted with the cream and sugar. That is what happens when we add more air to a burner with an elevated CO level. You are limited to 400 parts per million air free in a stack. To calculate that for natural gas using your O2 reading, use the following formula:

$$CO\ Air\ Free\ ppm = \left(\frac{20.9}{20.9 - O2}\right) X\ COppm$$

Let me show you a calculation where the CO reading is 50 ppm and the O2 is 4% and the CO2 is 10%

$$CO\ Air\ Free\ ppm = \left(\frac{20.9}{20.9 - 4}\right) X\ 50\ ppm\ CO$$

$$CO\ Air\ Free\ ppm = 62\ ppm$$

To calculate that for natural gas using your CO2 reading, use the following formula:

$$CO\ Air\ Free\ ppm = (\frac{12.2}{CO2})\ X\ COppm$$

$$CO\ Air\ Free\ ppm = (\frac{12.2}{10})\ X\ 50ppm$$

$$CO\ Air\ Free\ ppm\ 61 = 1.22\ X\ 50ppm$$

To calculate that for Propane your CO2 reading, use the following formula:

$$CO\ Air\ Free\ ppm = (\frac{14}{CO2})\ X\ COppm$$

Did you know that each cubic foot of natural gas burned produces:

- 8 cubic feet of nitrogen
- 2 cubic feet of water
- 1 cubic foot of carbon dioxide

Which Boiler is Safer?

Boiler 1	Boiler 2
Percent O2 3.5% As Measured CO Reading 300 ppm	Percent O2 18% As Measured CO Reading 75 ppm

Boiler 1 has a reading of 300 ppm carbon monoxide with 3.5% oxygen reading and Boiler 2 has a reading of 75 ppm carbon monoxide at 18% oxygen. While boiler 1 may seem worse than boiler 2, that is not the case. When you calculate the CO Air Free of each, boiler 1 has a reading of 360 ppm and boiler 2 has a reading of 541 ppm. While boiler 1 should be adjusted to try lowering the CO levels, it is below the 400 ppm air free threshold. Boiler 2 should not be operated in its current condition as it is dangerous and 35% higher than the 400 ppm CO Air Free threshold.

When checking the fuel to air ratios, it is common to see the carbon monoxide readings rise when the modulating burner changes firing rate. When setting the fuel to air ratio with the mechanical linkages, the CO levels will rise and then should return to the setting once the linkages align again. Allow the flame to settle at each of the firing rates you want to check before making adjustments.

Carbon monoxide in the flue gases could really make your day a challenging one as the CO level in the stack could go from zero to several thousand parts per million parts in seconds. If it does, many analyzers will shut down. To clear the carbon monoxide from the probe, the unit may go into a purge mode and you will not be able to use it. On severe cases, it may take several hours to clear the probe or in my case, the unit failed and had to be sent back to the factory. It cost me several hundred dollars. Really keep an eye on the carbon monoxide levels as you are adjusting the fuel to air ratio. If it starts to spike, I may find myself running to the rear of the boiler to pull the probe out or using a second technician to remove the probe to limit the exposure to the carbon monoxide. To limit this, I will adjust the gas pressure to the proper level and visually watch the flame before installing the probe. I like to carry a second analyzer with me during my service call in case the primary one fails.

Calculate CO Air Free from CO Readings							
Diluted or "As Measured" CO Reading							
O2%	**25**	**50**	**75**	**100**	**125**	**150**	**175**
1.0	26	53	79	105	131	158	184
2.0	28	55	83	111	138	166	194
3.0	29	58	88	117	146	175	204
4.0	31	62	93	124	155	186	216
5.0	33	66	99	131	164	197	230
6.0	35	70	105	140	175	210	245
7.0	38	75	113	150	188	226	263
8.0	41	81	122	162	203	243	284
9.0	44	88	132	176	220	263	307
10.0	48	96	144	192	240	288	336
11.0	52	106	158	211	264	317	369
12	59	117	176	235	394	352	411
13	66	132	198	265	331	397	463
14	76	151	227	303	379	454	530
15	89	177	266	354	443	531	620
O2%	**200**	**225**	**250**	**275**	**300**	**325**	**350**
1.0	210	236	263	289	315	341	368
2.0	221	249	276	304	332	359	387
3.0	234	263	292	321	350	379	409
4.0	247	278	309	340	371	402	433
5.0	263	296	329	361	394	427	460
6.0	281	316	351	386	421	456	491
7.0	301	338	376	413	451	489	526
8.0	324	365	405	446	486	527	567
9.0	351	395	439	483	527	571	615
10.0	383	431	479	527	575	623	671
11.0	422	475	528	581	633	686	739
12	470	528	587	646	704	763	822
13	529	595	661	728	794	860	926
14	606	682	757	833	909	984	1060
15	708	797	886	974	1063	1151	1240

Excess Air Excess air is the air that is not required for combustion. For example, natural gas typically requires ten parts of air for each part of gas for perfect efficiency. We cannot leave it at this point as the burner blower wheel could get dirty and affect the fuel to air ratio by not providing enough air. We will add a little more air to the fire to assure safe operation. Think of it like the secret service for the president. It is there in case something happens. We typically add about 2- 5 extra parts of air. That would equal 20-50% excess air.

Parts of Air per One Part of Gas	Excess Air Percentage
10	0%
11	10%
12	20%
13	30%
14	40%
15	50%
16	60%
17	70%
18	80%
19	90%
20	100%

Draft This is a measurement of how much draw or suction that we have in the chimney or stack. If the draft is excessive, it could pull the flue gases through the boiler too quickly and lose some of the efficiency of the boiler. Another issue is that excessive draft could cause the burner to over fire. If the manufacturer wants a definite gas pressure at the burner head, this is with a certain draft reading. If the draft is too high, the regulator will open further trying to reach the set pressure. If it is inadequate, we could damage the burner head because too much heat will stay at the burner. Draft is measured on boilers with a negative draw like Category 1 appliances. The draft is typically about -0.05" w.c. This should be verified with the manufacturer. It is measured with either the combustion analyzer or a manometer. Most condensing boilers have a positive flue and require a sealed flue.

Stack Temperature The stack temperature is a good indication of the heat transfer ability of the boiler. If the stack temperature is too high, this indicates that the burner is not transferring the heat into the boiler. The cause may be from a combination of dirty heating surfaces, scale buildup, excessive velocity of the flue gases or excessive velocity of the water inside the boiler. If the stack temperature is too low, the flue gases will condense and damage the boiler, flue, and chimney unless it is a boiler designed for condensing.

Ambient temperature The ambient temperature is the temperature of the combustion air for the burner. When calibrating your combustion analyzer, it should be done in the area where the combustion air is drawn, as close to the burner as possible. If you calibrate the probe outside and the boiler uses air from the boiler room, it will give a false reading. If your boiler uses direct vented combustion air, I like to calibrate my analyzer in the combustion air duct feeding the burner. If you have two boilers in the room and one takes outside air and one takes room air, you will have to calibrate your analyzer twice; once for each type of burner.

Smoke spot If the boiler uses oil as the fuel, you will need to test for smoke inside the flue with a smoke tester. Smoke in the flue gases means incomplete combustion. The smoke tester pulls flue gases across a piece of white paper. When done, you compare the color on the sheet to a smoke scale that will tell you the amount of smoke in the flue. If the smoke is excessive, the burner requires additional service.

Always, Always get a print out or picture Whenever you test the combustion on a jobsite, always get a printout of the readings. This is meant to protect you and your company. If something were to fail, you would have the printout of your findings with a time and date stamp. When using a thermal printout, they will sometimes fade so I will make a copy of the sheet and place it in the customer file. If you have a cell phone with a camera, you could take a picture of the readings and save the picture for your records. Legal suits, both civilly and criminally, are becoming more common place for service technicians. You have to protect yourself by using documentation.

Leave a copy in the burner cabinet When I am done with the adjustments, I leave a copy of my readings in the burner control panel. This is an excellent reference source for the next service call. You can compare current readings with the startup readings and see what adjustments should be made.

What efficiency should we adjust the burner for? When starting a boiler, I reference the stated efficiency of the boiler and feel that if I am within one percent of the published efficiency, I will be satisfied with that. When a new boiler is tested by an independent organization for its published efficiency, it is under optimum conditions. In reality, the conditions on the job are not the same as the one at the testing lab.

Combustion Adjustment We had a new client that was referred to us. The building owner had to rebuild the refractory and replace several boiler tubes each year. When we checked the combustion efficiency, the boiler was reading about 88% efficient at high fire and almost 90% efficient at low fire. The boiler was rated for 80% efficient from the manufacturer. The previous service technician had reduced the firing rate below the condensing point. The acids in the flue gases were destroying the boiler when they condensed. We explained the problem to the owner and suggested that we adjust the efficiency to the proper settings. The owner asked us to keep the settings as they were and he said that he would rather sacrifice some tubes than to increase his fuel costs. I told him that I could not purposely operate the boiler at dangerous condition and he chose not to do business with us and I was ok with that.

Combustion Analysis Misc. Items

- Do not forget to seal the hole for the combustion analysis probe! This is critical on boilers with a positive flue as the flue gases could be pushed out the hole.
- Mark your setting on linkages. This may help in the future if the linkages slip. A permanent marker is good for this.

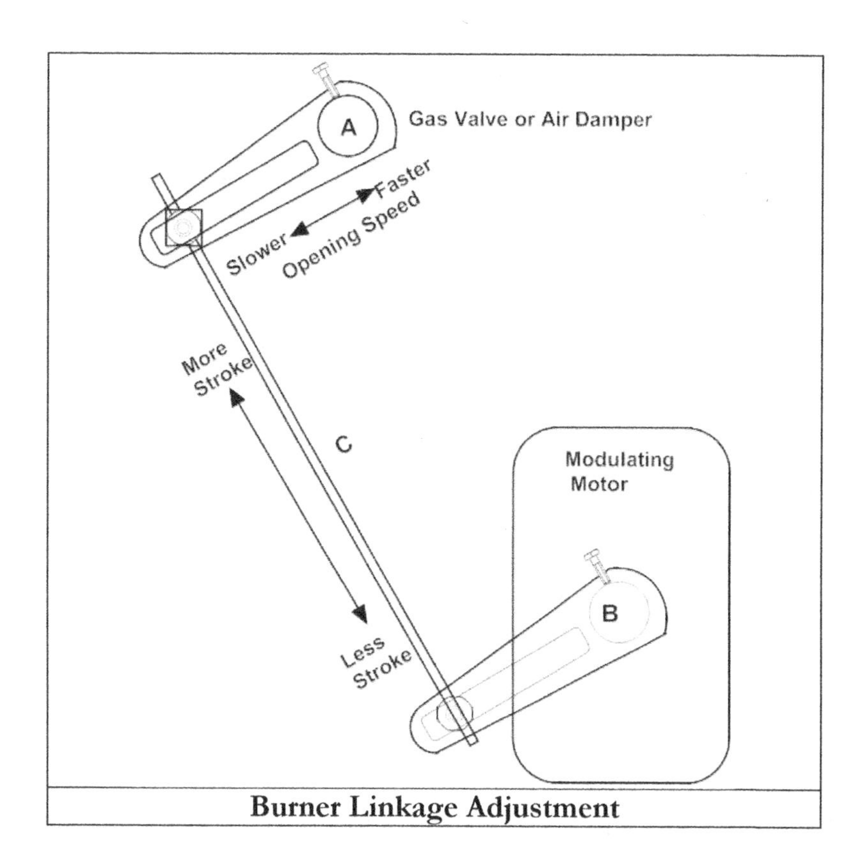

Burner Linkage Adjustment

Burner Linkage Adjustment When adjusting the fuel to air ratio, you will have to adjust the linkages on the burner and combustion air. The above picture may help when adjusting the fuel to air to know which way to move it. For instance, if the burner needed more air, you could move the top adjustment closer to the A mark as it will open it faster.

Combustion Analysis A modulating burner will take much longer to check and adjust than an on-off or low high low burner which increases the maintenance costs. In addition, it is almost impossible to get the same efficiency throughout the entire range. The new linkage-less controls make the setup somewhat easier. The linkage less controls are typically used on boilers larger than 150 boiler horsepower.

```
           testo 327-1
V1.18           01909540/USA

12/20/2012          09:44:31

Fuel                 Nat Gas
CO2 max               11.7 %

          Flue gas

 428.7 °F      T stack
  8.91 %       CO2
  82.9 %       EFF
  27.2 %       ExAir
   4.9 %       Oxygen
     3 ppm     CO
     4 ppm     CO AirFree
   ---   inH2O Draft
  68.4 °F      Ambient temp
  60.1 °F      Instrum temp
   --- °F      Diff. temp.
   ---   inH2O Diff. Press
     0 ppm     CO Ambient
```

When adjusting the fuel to air ratio at low fire, a subtle change has taken place in the industry. Many of the burner manufacturers have found that when the burners stay on low fire for extended times, damage to the burner heads may occur. In addition, the fuel and air do not mix as well, leading to elevated carbon monoxide and emission levels. To get better mixing and better burner protection, you will now see that many burners require more air at low fire. In some instances, it may be 20-40% more at low fire. This could lower the efficiency by 1-2% more. In addition, extended operation at low fire in standard boilers could allow the flue gases to condense, destroying the boiler with category I vents.

There are many things that affect the fuel to air ratio of the burner. It could be something as small as cleaning the blower wheel. When cleaned, it brings more air to the burner and combustion process, affecting the fuel to air ratio. When working on a boiler, it is a good idea to check the combustion during every service call. The linkages could have slipped or changed since the last visit. It is crucial if you change or adjust the following components: gas valve, blower motor, blower wheel, gas pressure.

Typical Combustion Test Results

Category 1 Boilers

Atmospheric Gas Burner

Oxygen	7% - 9%
Stack Temperature	325 to 500 Degrees F
Draft in Water Column Inches	-.0 " WC to -.04" WC
Carbon Monoxide PPM	<100 PPM Air Free

Gas Power Burner

Oxygen	3% - 6%
Stack Temperature	275 to 500 Degrees F
Draft in Water Column Inches	-.02" WC to -.04" WC
Carbon Monoxide PPM	<100 PPM Air Free

Oil Power Burner

Oxygen	4% - 7%
Stack Temperature	325 to 600 Degrees F
Draft in Water Column Inches	-.04" WC to -.06" WC
Carbon Monoxide PPM	<100 PPM Air Free with 0 Smoke

Estimated Combustion Efficiency Tables Less than 90% Efficient

Natural Gas

Excess Air	O2%	CO2%	Net Stack Temperature						
			200	250	300	350	400	450	500
8.5	2.0	10.7	85.4	84.2	83.1	81.9	80.8	79.6	78.4
12.1	2.5	10.4	85.3	84.1	83.0	81.8	80.3	79.4	78.2
15.0	3.0	10.1	85.2	84.0	82.8	81.6	80.1	79.2	77.9
18.0	3.5	9.8	85.1	83.9	82.6	81.4	79.9	78.9	77.6
21.1	4.0	9.6	85.0	83.7	82.5	81.2	79.7	78.7	77.4
24.5	4.5	9.3	84.8	83.6	82.3	81.0	79.4	78.4	77.1
28.1	5.0	9.0	84.7	83.4	82.1	80.8	79.2	78.1	76.7
31.9	5.5	8.7	84.6	83.3	81.9	80.6	78.9	77.8	76.4
35.9	6.0	8.4	84.4	83.1	81.7	80.3	78.6	77.5	76.0
40.3	6.5	8.2	84.3	82.9	81.5	80.0	78.3	77.1	75.6
44.9	7.0	7.9	84.1	82.7	81.2	79.7	77.9	76.7	75.2
49.9	7.5	7.6	84.0	82.5	80.9	79.4	77.6	76.3	74.8
55.3	8.0	7.3	83.8	82.2	80.7	79.1	77.2	75.9	74.3

Estimated Combustion Efficiency Tables Less than 90% Efficient

#2 Fuel Oil

Excess Air	O2%	CO2%	Net Stack Temperature						
			200	250	300	350	400	450	500
9.9	2.0	14.1	89.6	88.5	87.4	86.3	85.2	84.1	82.9
12.6	2.5	13.8	89.5	88.4	87.3	86.2	85.0	83.9	82.7
15.6	3.0	13.4	89.4	88.3	87.1	86.0	84.8	83.6	82.4
18.7	3.5	13.0	89.3	88.2	87.0	85.8	84.6	83.4	82.2
22.0	4.0	12.6	89.2	88.0	86.8	85.6	84.4	83.1	81.9
25.5	4.5	12.3	89.1	87.9	86.6	85.4	84.1	82.8	81.6
29.2	5.0	11.9	89.0	87.7	86.4	85.1	83.9	82.6	81.2
33.2	5.5	11.5	88.8	87.5	86.2	84.9	83.6	82.2	80.9
37.4	6.0	11.2	88.7	87.3	86.0	84.6	83.3	81.9	80.5
41.9	6.5	10.8	88.5	87.1	85.8	84.4	83.0	81.5	80.1
46.8	7.0	10.4	88.3	86.9	85.5	84.1	82.6	81.2	79.7
52.0	7.5	10.0	88.2	86.7	85.2	83.8	82.3	80.7	79.2
57.6	8.0	9.7	88.0	86.5	84.9	83.4	81.9	80.3	78.7

Pollution Conversions After you finish the combustion analysis on the boiler, you may be asked to forward an emissions report. To calculate that, take the readings and follow the following formulas. To convert from PPM to any of the units below, multiply PPM by the number in the correct column and row

Definitions

Lb/MBTU = pounds of pollutants per million BTU

Mg/NM3 = Milligrams of pollutants per Million BTU

	Multiply PPM by factor below				
Fuel	**Pollutant**	**Lb/MBTU**	**MG/NM3**	**MG/KG**	**G/GJ**
Nat Gas	CO	0.00078	1.249	12.647	0.338
Nat Gas	NOx	0.00129	2.053	20.788	0.556
Nat Gas	SO2	0.00179	2.857	28.949	0.775
Oil #2, #6	CO	0.00081	1.249	15.118	0.354
Oil #2, #6	NOx	0.00134	2.053	24.850	0.582
Oil #2, #6	SO2	0.00186	2.857	34.605	0.811

MG/KG = Milligrams of pollutants per Kilogram of fuel burned

G/GJ = Grams of pollutant per Gigajoule

Based on 3% excess Oxygen and dry gas

Boiler Emissions Due to worries about global climate change, boiler emissions are a growing concern. Many states are requiring reduced emissions from the boilers. Please check with your state to determine their emission limits. In some states, different areas or cities may have different requirements. There are several flue gas by-products that are being regulated.

Nitrogen Compounds or NO_x NO_x is principally made up of two components, Nitric Oxide(NO) and Nitrogen Dioxide(NO_2). NO_2, when combined with other pollutants, such as Volatile Organic Compounds (VOCs), is believed to form O3 or ground level ozone and acid rain. NO_x emissions become more prevalent when the flame temperature is above 2,800^{0}F and the fuel to air combustion ratios are between 5-7% O_2. This is typically where the commercial burner manufacturers direct the fuel to air ratios to be set.

There are two types of NOx, Thermal NOx and Fuel NOx. Thermal NOx is the most common. It is produced in the boiler when oxygen and nitrogen combine under elevated temperatures. Fuel NOx rarely occurs when firing with gas. It is common in heavier fuel oils.

Particulate Matter Particulate matter, or otherwise known as soot, is mostly formed from incomplete combustion of fuel. It is comprised of unburned fuel, organic chemicals, soil, dust, sulfates, nitrates, oxides and or carbons. There are two basic types of particulate matter, Pm and PM_{10}. PM_{10} is particulate matter that is less that 10 microns in diameter. To see how small this is, consider that a human hair is about 70 microns. The PM_{10} is small enough to bypass the human body's natural filtering system and imbed themselves in the lungs. It can trigger asthma attacks, coughing and acute bronchitis. It is more prevalent in oil than gas. There are some areas that are trying to test for particulates as small as 2.5 microns.

Sulfur Dioxide Sulfur Dioxide is also a by-product of combustion. It is rare with gaseous fuels. It is also a primary contributor to acid rain, which causes acidification of streams and lakes. It is released primarily from burning fuels that contain sulfur (such as coal, oil, and diesel fuel). Sulfur dioxide, when combined with water, forms sulfuric acid.

Carbon Dioxide Carbon dioxide, when combined with water, forms carbonic acid.

Acid Rain and Stack Temperature The flue gases from a fossil-fueled boiler contain the following: oxygen, carbon dioxide, carbon monoxide, sulfur dioxide and free water. If allowed to condense, these acids will destroy the stack, chimney and boiler. The following are the acid dew point temperatures of the most common fuels. Flue gas condensation could also occur if the burner is under fired. This means that the supply of fuel to the burner is less than the manufacturer's recommendations.

Fuel	Acid Dew Point Temperature	Minimum Allowable Stack Temperature
Natural Gas	150	250
#2 Fuel Oil	180	275

Did you know that natural gas is actually odorless and colorless? Mercaptan is added to the gas to give it that distinctive rotten egg smell.

Combustion Air Required For Boilers

Atmospheric with Draft Hood	
Cubic Feet Air per Cubic Foot Gas	Description
10	Perfect Combustion
5	Primary Air (50% excess air)
15	Air for Draft Hood (Secondary Air)
30	Total Air Required

Boiler With Power Burner & Barometric Damper	
Cubic Feet Air per Cubic Foot Gas	Description
10	Perfect Combustion
2	Primary Air (20% excess air)
3	Air for Barometric Damper (Secondary Air)
15	Total Air Required

Boiler With Power Burner & NO Barometric Damper	
Cubic Feet Air per Cubic Foot Gas	Description
10	Perfect Combustion
2	Primary Air (20% excess air)
12	Total Air Required

A Note on Direct Vent Boilers When you are adjusting the fuel to air ratio on a burner with the combustion air directly vented from the outside, you should be aware of the outside temperature at which you are working. If the weather is very cold when you are adjusting the fuel to air ratio, please be aware that you will have less air when the weather gets warmer due to the air's density.

How combustion air affects boiler efficiency When you are adjusting the fuel to air ratio of a burner, the amount of excess air can directly affect the dew point or condensing temperature of the flue gases. For example, we are taught that flue gases will condense at 140 degrees F but that only happens when the oxygen content of the flue gases is below 3%. If we have 5% oxygen content in the flue gases, our condensing temperature drops 10 degrees to 130 degrees F. The chart below will show the dew point temperatures of flue gases at various oxygen levels. At the 20% excess air that many burners use, the condensing temperature of the boiler is now 131^{0}F. That is the design temperature for European condensing hydronic systems.

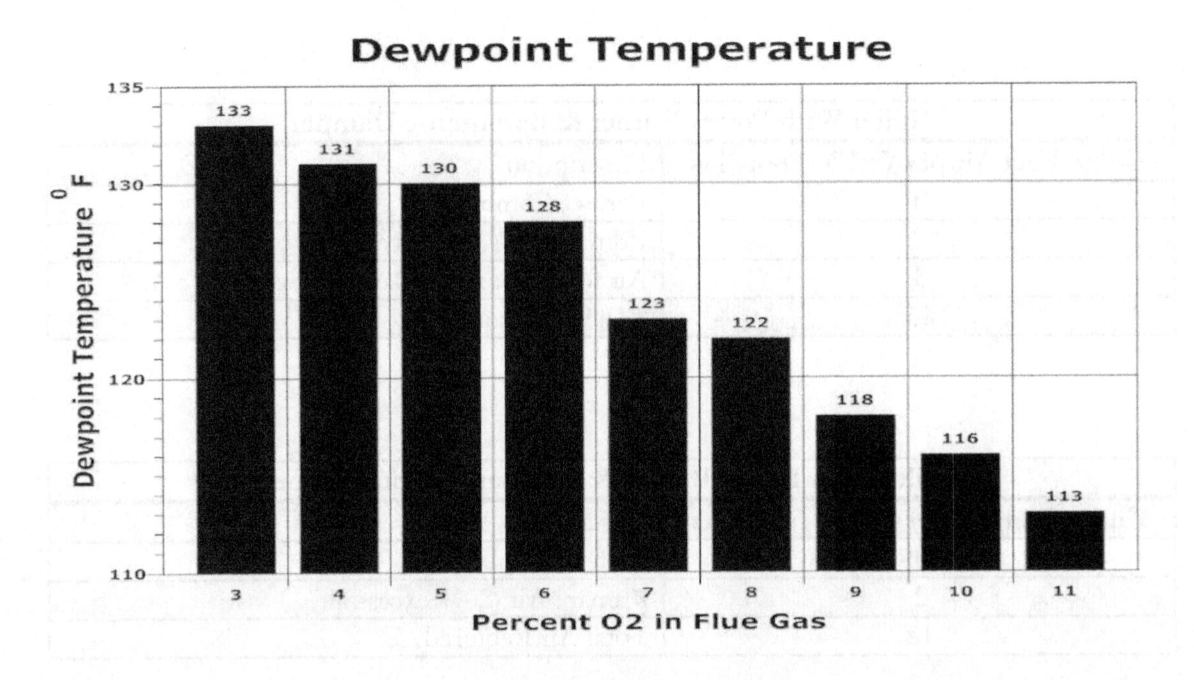

Flue Gas Dew Point Temperature			
O2 %	CO2 %	Excess Air %	Dew Point ^{0}F
3.0	**10.0**	**15.0**	**133**
4.0	**9.5**	**20.0**	**131**
5.0	**9.0**	**29.0**	**130**
6.0	8.4	**36.0**	**128**
7.0	**7.9**	**46.5**	**123**
8.0	7.3	**56.5**	**122**
9.0	**6.7**	**68.6**	**118**
10.0	6.2	83.5	**116**
11.0	**5.6**	**100.0**	**113**

Rule of thumb. For every 0.01" w.c. the excess draft can be reduced, the fuel consumption is reduced by 1% in Category 1 appliances. On Category 4 boilers, the draft is about 1" positive.

Typical Draft Readings for Boilers		
Type of Heating System	**Overfire Draft**	**Stack Draft**
Gas, Atmospheric	Not Applicable	-.02 to -.04" WC
Gas, Power Burner	-.02" WC	-.02 to -.04" WC
Oil, Conventional	-.02" WC	-.04 to -.06" WC
Oil, Flame Retention	-.02" WC	-.04 to -.06" WC
Positive Overfire Oil & Gas	+.4 to +.6	-.02 to -.04" WC
Category 4 Positive	Positive	+1.0" w.c.

The table below shows the effect of temperature and barometric conditions on excess air percentages in the burner if the burner was set up at 80 degree F combustion air, 15% excess air and a barometric pressure of 29.

Effect of Air Temperature at Same Barometric Pressure		
Combustion Air Temperature	Barometric Pressure	Excess Air %
40	29	25.5
60	29	20.2
80*	**29***	**15***
100	29	9.6
120	29	1.1
Effect of Barometric Changes at Same Temperature		
Combustion Air Temperature	Barometric Pressure	Excess Air %
80	27	7
80	28	11
80*	**29***	**15***
80	30	19
Effect of Temperature and Barometric Changes		
Combustion Air Temperature	Barometric Pressure	Excess Air %
40	31	34.5
60	30	25
80*	**29***	**15***
100	28	5
120	27	-5.5
*Burner Setup Conditions = 80 Degrees ambient air, 15% Excess Air & 29 Barometric Pressure		

What happens when we adjust the combustion efficiency when the weather is cold? I was taught that the proper time to adjust the combustion efficiency of a burner is when the temperature is cold and the boiler can operate longer. See what happens when we adjust the fuel to air ratio when it is 0 degrees F. The burner will be operating in a dangerous condition when the outdoor temperature rises to even 20 degrees F.

Effect of Air Temperature on Boiler Efficiency		
Combustion Air Temperature	Barometric Pressure	Excess Air %
0	29	15%
20*	29	9.8%
40*	29	4.6%
60*	29	-0.6%
80*	29	-5.8%
Perform combustion adjustment at 0 degrees outside *Boiler will be operating in dangerous condition		

Percent of Air that will permit combustion

	Minimum % of Air	Maximum % of Air
Natural Gas	64	247
Oil	30	173
Coal Pulverized	8	425

Did you know that the air we breathe is about 21% oxygen?

Boiler Soot

Boiler soot is very dangerous if you see it on the boiler. Soot looks like fine black dust on the fireside of the boiler. It contains fine particulate matter (PM) that could embed itself in your lungs. It may be caused by incomplete combustion, low flue gas temperature, a leak in the boiler, inadequate combustion air, or a malfunction of the burner. The soot is highly flammable and extreme caution should be used if encountered. It is also a real mess to clean as it goes everywhere. In addition, soot insulated the fireside of the boiler, reducing the efficiency.

20 Questions We used to service the boilers at a high school in my area and the custodian for the school would ask questions the whole time we were there. We jokingly referred to him as 20 Questions. He wanted to know everything about the boilers. Due to a vacancy in the maintenance department, the custodian bid on and received the maintenance position. He was a hard worker and a conscientious employee. One day, the boilers sooted and the custodian never called to ask about the proper way to clean the boiler. He used some of the old wire brushes that were in the maintenance department. The soot went everywhere and covered the man. On one stroke of the metal brush, it sparked and the soot ignited. It burned inside the boiler and also caught the man's uniform on fire. He received 3rd degree burns on his arms and 2nd degree ones on his face. He was never able to work again. Use caution when cleaning a sooted boiler.

Cleaning a Sooted Boiler To remove the soot, you will need to be covered from head to toe. Breathing masks with fine filters should be used as the soot particulate is very fine. All skin should be covered because the soot will not be removed with plain soap and water. It will have wear itself off. I made the mistake of cleaning a sooted boiler with loose fitting clothing the day of my wife's work Christmas party. I looked like one of those old coal miners from a century ago. I tried everything to remove it including cleanser, auto mechanic soap, pot scouring pads and even my wife's cold cream and nothing worked. The good part was that I got out of attending her Christmas party but paid for missing it ten times over.

Once you are suited up, you want to think about the tools you will use for cleaning the boiler. I like to get a spray bottle filled with water and carefully spray the soot to make it less friable and easier to clean. If you are using water to clean the soot, the soot could turn acidic . Use caution not to wet the refractory inside the boiler as the water could damage or ruin it. I use Nylon or plastic brushes and a large shop vacuum with a fine filter inside to catch the soot. A technician in a seminar once told me that he used compressed air to clean a sooted boiler and offered that it was one of his worst decision. The soot went everywhere and the customer made him clean the entire boiler room and took several days to clean it. In addition, you want to be careful that it does not enter anywhere where it can ignite, such as electrical panels or water heater. During one of my seminars, an old navy guy told me his secret of how to stay "soot free" when cleaning a sooted boiler. He informed me that he would have to clean the boilers every month because the water temperature was too low. His attire for cleaning the boiler consisted of sealing the sleeves of his shirt with duct tape and covering his face with Vaseline to keep the soot from imbedding itself in his skin. He would wear a tight fitting hat and coveralls. When I asked why he simply did not raise the water temperature, he told me his boss wanted to save money. If the boiler is a cast iron boiler, it is very difficult to clean the fireside of the boiler due to the narrow flue passages. In some severe instances, the sections may need to be disassembled to properly clean the boiler. It is always less expensive to verify that the burner is operating correctly than to allow it to soot the boiler. Another consideration with boiler soot is that it will float on the water when you are cleaning it so use caution when hosing the floor when cleaning.

Effect of Soot on Fuel Consumption Soot accumulation on the heating surfaces can dramatically increase fuel consumption.
Please see below.

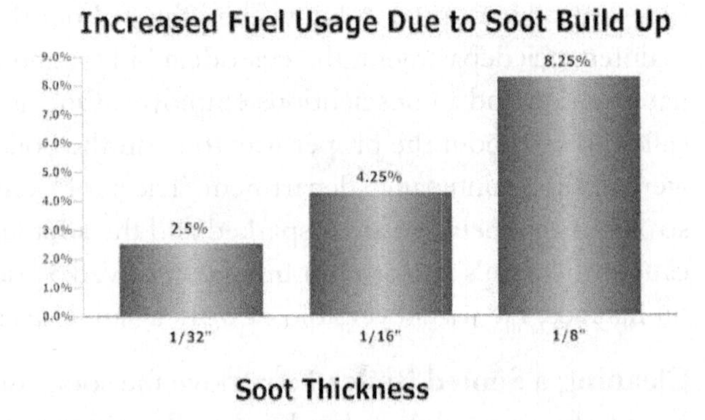

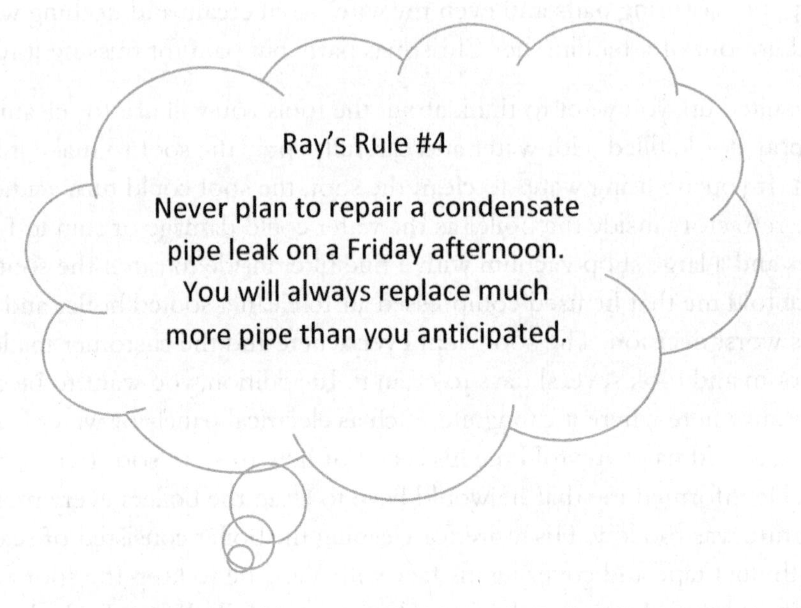

Typical Sequence for On Off Power Burners

1. On a call for heat, the blower motor starts and is verified with the burner airflow switch.
2. Burner starts pre-purge sequence and this could be from 15 seconds to 2 minutes.
3. After pre-purge, burner will light the pilot flame and verify the presence of a flame.
4. Once pilot is verified, the flame safeguard will supply electricity to safety shutoff valves, opening them. The flame signal will be tested by the flame safeguard.
5. The burner stays on until the call for heat has ended.

Typical Sequence for Low High Off Power Burners

1. On a call for heat, the blower motor starts and is verified with the burner airflow switch.
2. Burner goes into pre-purge sequence and this could be from 15 seconds to 2 minutes.
3. After pre-purge, burner will light the pilot flame and verify the presence of a flame.
4. When pilot is verified, the flame safeguard will supply electricity to safety shutoff valves, opening them. The flame signal will be tested by the flame safeguard.
5. Once flame is established, the burner drives to high fire and will stay there until the call for heat has ended.

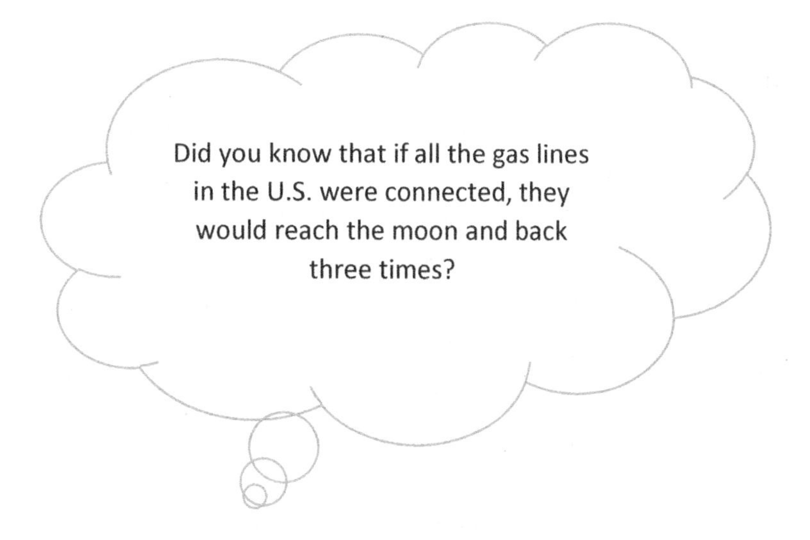

Typical Sequence for Low High Low Power Burners

1. On a call for heat, the blower motor starts and is verified with the burner airflow switch.
2. Burner goes into pre-purge sequence and this could be from 15 seconds to 2 minutes.
3. After pre-purge, burner will light the pilot flame and verify the presence of a flame.
4. When pilot is verified, the flame safeguard will supply electricity to safety shutoff valves, opening them. The flame signal will be tested by the flame safeguard.
5. After flame is established, the burner drives to high fire to meet the setpoint of the firing rate control. Once it reaches the setpoint of the firing rate control, it will then drop to low fire.
6. Once it meets the setpoint of the firing rate control, it will stay at low fire until the call for heat has ended or the boiler temperature or pressure falls below the setpoint of the firing rate control. The burner will travel between low and high fire until the call for heat has ended.

Typical Sequence for Modulating Power Burners

1. On a call for heat, the blower motor starts and is verified with the burner airflow switch.
2. Burner goes into pre-purge sequence and this could be from 15 seconds to 2 minutes.
3. During pre-purge, the modulating motor may drive to the high fire position. Depending on the burner, there may be an internal switch in the modulating motor that will be made at the high fire position. At the end of the pre-purge cycle, the burner will drive to the low fire position and may make an internal switch inside the modulating motor. This switch is called the *Low Fire Start Switch.*
4. After pre-purge, burner will light the pilot flame and verify the presence of a flame.
5. After pilot is verified, the flame safeguard will supply electricity to safety shutoff valves, opening them. The flame signal will be tested by the flame safeguard.
6. Once flame is established, the burner drives to high fire to meet the setpoint of the firing rate control. As it gets close to the setpoint, the firing rate or modulating control will send a signal to the modulating motor to drive it between low and high fire in an attempt to stay at the setpoint.
7. The burner will travel between low and high fire until the call for heat has ended.

Water Treatment

You see a nipple leaking on the condensate line on a Friday afternoon and think that you can replace that leaking nipple in only a few minutes. You place a pipe wrench on the pipe and get that sick feeling in your stomach as the nipple collapses. As you inspect the nipple, you see that it is paper-thin. If one nipple is paper thin, the adjoining nipples and pipe are probably in the same condition. You typically have to go back to the closest fitting and replace the piping. It is not something that I would attempt on a Friday afternoon. Leaks on steam piping are usually due to two causes, oxygen pitting and carbonic acid. Oxygen pitting will be selective and only attack certain areas of the pipe and you will see pin holes in the pipe. Carbonic acid is formed when carbon dioxide mixes with water and this will erode the bottoms of the pipe. Remember Ray's Rule *4 Never plan to repair a condensate pipe leak on a Friday afternoon.*

Scale Build Up Scale formation on the waterside of the boiler could dramatically increase the operating costs. The chart to the right illustrates the estimated costs of scale formation. Scale accumulates from an increase in makeup water due to a leak in the system or if the building has hard water. It will be drawn to the hottest surface. Scale buildup sounds like Rice Kripies Cereal i.e. Snap, Crackle, Pop when the flame is on.

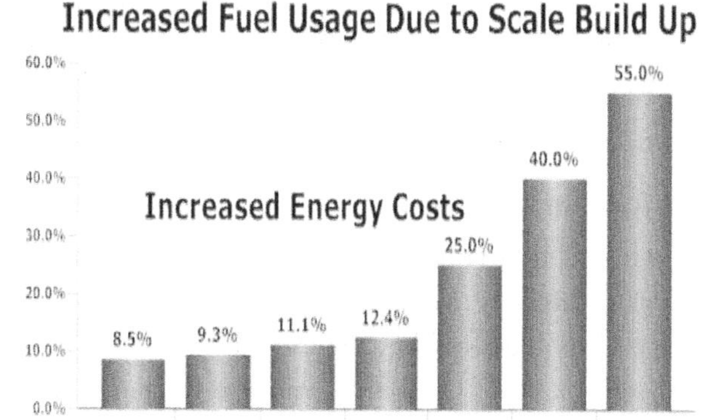

Water Treatment Most boilers require some sort of water treatment. A hydronic system and a steam system may use different types of chemical feed systems.

Pot feeders (See Picture) are traditionally used on hydronic systems. The pot feeder is usually a side-arm type feeder that has isolation valves. The pot feeder could also include a filter that will strain the system water. If using a filter as well, a flow indicator should be installed to indicate whether the filter is plugged and should be cleaned or changed.

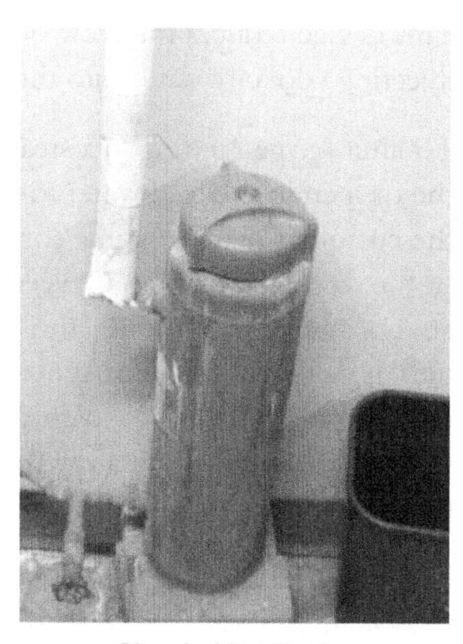
Chemical Pot Feeder

On steam systems, several chemicals may be introduced into the system. Some may be injected into the boiler feed unit and some may be directly injected into the steam supply piping. Makeup Water, Oxygen and Carbon Dioxide are the three main enemies of a steam system. Makeup water introduces hardness, oxygen and carbon dioxide into the system. Oxygen molecules will pit the inside of the boiler or piping causing a leak. Carbon Dioxide forms Carbonic Acid inside the steam system. Carbonic acid occurs when carbon dioxide is mixed with water. This is usually noticed on the condensate piping. A way to reduce carbonic acid is to insulate the condensate piping as this will reduce carbon dioxide levels inside the pipe.

Boiler Blowdown Tank

The most effective water treatment is a tight system that has no leaks. Makeup water will bring in fresh, untreated water. This fresh water contains minerals such as calcium and magnesium. This is usually referred to as "Hardness" These minerals will deposit themselves on the hottest surface and act as an insulator. This reduces the heat transfer ability of the boiler as well as the efficiency. In severe instances, it could destroy the boiler. Hardness inside a steam boiler should be limited. One way to limit the hardness is to install a water softener. A softener also reduces the amount of water treatment chemicals that are used.

Blow down is crucial on steam systems and provisions should be made for a blow down valve and / or a surface skim tapping. Location of the blow down fitting is critical as dissolved solids will generally collect at 4-6" below the water level. The skim tapping is typically just below the normal boiler operating level. Some systems require automatic blow down of the boilers. In this case, the water treatment expert could inform you of this. Also, check your installation codes to see if the blow down piping can be piped directly to the drain. Many municipalities do not allow that because the high temperature boiler water could damage the PVC drainpipes in the building. In that case, you would need a blow down cooler. This device reduces the blow down discharge temperature to allow safe draining of the water by injecting cool city water into the unit.

To limit oxygen pitting in a steam system, a deaerator is often installed. This will remove most of the oxygen in the feed water to the boiler, limiting the damage from oxygen pitting. In addition, the customer may use some sort of oxygen scavenger water treatment chemical to remove more of the dissolved oxygen. Oxygen pitting is an amazing process. If you look at a nipple that was attacked by oxygen pitting, the oxygen molecules will attack the same spot over and over again until it develops a hole. It almost looks like someone drilled a hole in the nipple. The rest of the nipple could be relatively damage free. It is like a shark that is attracted to the scent of blood. Carbonic acid will wash away the bottom of the nipple or piping. The top of the piping may be the same thickness as the day it was installed.

Should dye be used in the water treatment? Colored dye is sometimes included with the water treatment as a way to visually display if the system piping is leaking. One hospital in our area found the value of the dye in their treatment. They had a stainless steel water fountain installed when renovating one of their wings. The fountain was in service for several months without incident until one day, a nurse filled a white Styrofoam cup with water from the fountain. She saw that the water had a pink hue. The nurse called the maintenance department and asked why the water was pink. After some investigation, it was discovered that the installer had connected the fountain to the chilled water loop. Without the dye, the mistake could have never been found. Many people could have unknowingly ingested the water treatment chemicals.

Feeding the Chemical Treatment Chemical treatment for steam systems should be proportionately fed into a heating system. Typically, the water treatment company may use an injection pump for each chemical. The pump feeds a small amount of chemical at a time. Some facilities will manually feed the chemical into the boiler feed tank if their reading is low. This is referred to as "slug feeding" There are several drawbacks to this type of chemical feeding. First of all, the employees would be subjected to having the chemical spill on them when pouring it into the boiler feed tank. In addition, the system will experience wide variations of the chemical levels inside the boilers. This could lead to bouncing water levels and potential boiler problems. The other disadvantage is that there will be a disproportionate amount of chemicals in each boiler. The lead boiler on the day that the chemicals are fed will receive most of the chemicals. The lag boilers will get a substantial amount less or even none at all.

Chemical Feed Pump

Water Treatment Terms

Carry-Over: It is the continual discharge of impurities with steam. It can be detected by checking the conductivity of the condensate water and comparing it to the boiler conductivity.

Caustic Embrittlement: It is the weakening of the steel because of inner crystalline cracks. It is caused by long exposure to highly alkaline water and or stress of the metal.

Corrosion: It is the result of low-alkaline boiler water or the presence of free oxygen or both.

Foaming: It is a layer of foam on the surface of the water. It is most commonly caused by oil and other impurities. It may appear as a film atop the water, which impedes the steam bubbles from breaking through.

Priming: It is when large slugs of water are suddenly discharged from the boiler with the steam. This is caused by impurities in the water and boiler design.

Scale: It is a deposit of solids that form on the heating surfaces.

TDS: Total Dissolved Solids

Typical Steam Boiler Water Treatment Normal Levels

TDS 1,500-3,000 ppm or 2,000-4,000 micromhos or microsiemens

Microsiemens x .70 = TDS

Phosphate 30-60 ppm

Hydroxyl alkalinity 200-400 ppm

Sulfite 30-60 ppm SO3

Boiler pH 7-9* Or 9.5-11(*pH over 11 can cause boiler foaming in some boilers.*)

Some boilers prefer 10-11.0 pH

Condensate pH 8.2-9.0

**Each boiler is different. Please check with manufacturer as to their requirements. The above are just some industry rules of thumb.*

Water Treatment Equations

$$\text{Cycles of Concentration} = \frac{Chlorides\ or\ TDS\ in\ Boiler\ Water}{Chlorides\ or\ TDS\ in\ Feed\ water}$$

Boiler Blow down = 4 to 8 % of boiler makeup

One Cubic foot of water = 7.5 Gallons

$$\text{\% of Blowdown} = \frac{Chlorides\ or\ TDS\ in\ Feedwater}{Chlorides\ or\ TDS\ in\ Boiler\ Water}$$

$$\text{\% of Makeup} = \frac{Chlorides\ or\ TDS\ in\ Feedwater}{Chlorides\ or\ TDS\ in\ Make\text{-}Up\ Water}$$

$$\text{Pounds of Chemical/1,000 Gallons Water} = \frac{PPM\ of\ Product}{120*}$$

*120,000 gallons of water weighs about 1,000,000 pounds.

Skimming the steam boiler Most steam boilers have a skim fitting which is located near the water line so that the solids can be skimmed from the boiler. This will help to keep the boilers running efficiently. If piping work has been done on steam system, the water level is where the cutting oil from cutting threads will end up. This could cause bouncing of water level. You may be able to remove the oil through the skim tapping.

A Note on pH A quality pH tester is a good service tool when diagnosing steam boilers as well as some of the newer condensing boilers. Each boiler manufacturer will require different pH levels inside their boiler. For instance, most steam boilers require a pH level of 7-9 while some manufacturers prefer 10-11.5 Levels that high could cause foaming and bouncing water levels in the boiler designed for 7-9. What some people do not realize is that pH readings are logarithmic. For instance, a reading of 8 is **Ten** times more acidic than a reading of 7 and a reading of 9 is **Ten** times more acidic than a reading of 8. A reading of 9 is **One Hundred** times more acidic than a reading of 7. That is why a proper pH reading is critical on boilers.

pH Detector

What happens when the pH is not correct? If the pH is too low it could cause acids to form which will attack the metal surfaces. If the pH is too high, it could cause an alkaline condition which will allow scale to form. It could also cause caustic embrittlement. If the pH is above 8, copper forms a copper oxide film which protects the metal.

pH Scale

pH	Equivalent	If pH was measured with dollars	
PH = 0	Battery Acid	-$10,000,000	
PH = 1	Hydrochloric Acid	-$1,000,000	
PH = 2	Lemon Juice, Vinegar	-$100,000	
PH = 3	Grapefruit, Orange Juice	-$10,000	Acidic
PH = 4	Acid Rain, Tomato Juice	-$1,000	
PH = 5	Black Coffee	-$100	
PH = 6	Urine, Saliva	-$10	
PH = 7	***"Pure" Water***	***$1***	***Neutral***
PH = 8	Sea Water	$10	
PH = 9	Baking Soda	$100	
PH = 10	Milk of Magnesia	$1,000	
PH = 11	Ammonia Solution	$10,000	Alkaline
PH = 12	Soapy Water	$100,000	
PH = 13	Bleaches, Oven Cleaner	$1,000,000	
PH = 14	Liquid Drain Cleaner	$10,000,000	

Flooded Boiler Room

In your career, you may find yourself in a boiler room that was in a flood. The flood could have been caused by nature or by a mechanical failure. If you are the person checking the system after a flood, use extreme caution. My feeling is that any component that was sprayed with or under water should be replaced. A building we looked at decided to dry his components and reuse them after I cautioned the customer to not do so. A couple months later, the flame safeguard failed and the electric gas shutoff started leaking gas into the boiler. There was a loud explosion that blew the doors off the boiler.

According to Weil McLain technical service bulletin no: SB-1205, salt water is much worse for a boiler. This is their statement, "***Therefore, Weil-McLain equipment contaminated with saltwater or polluted water will no longer be covered under warranty and should be replaced."*** It sounds serious to me. If you choose to re-use the boiler, Weil McLain suggests the following:

- *Replace all controls, gas valves, and electrical wiring on the boiler.*
- *Thoroughly inspect all burner tubes, gas piping, manifolds, orifices, and flue ways for signs of rust and/or sediment from the flood waters.*
- *For oil-fired boilers, replace all oil burners.*
- *Replace all insulation that has become water damaged.*
- *Where possible, inspect seal rings for damage from petroleum products.*
- *Thoroughly inspect all venting for signs of corrosion.*

Burnham also has a strong statement about boilers that have been in a flood, *"Burnham Commercial strongly recommends that any boiler that has been exposed to flood water be removed and replaced with a new boiler." This was in a letter dated November 5, 2012.*

You should consult the equipment manufacturer before choosing to re-use the flooded boiler. I would not like to take the responsibility for telling the client to continue using the boiler.

Hydronic Boilers

Water temperature Verify with the boiler manufacturer to see what is the lowest recommended water temperature. Most traditional hydronic boilers were designed to operate at 180^{0}F and will condense if the return water temperature is below 140^0 F. The condensing boiler is designed to operate efficiently at temperatures below 140^{0}F and the efficiency improves the colder the water is.

Delta T or Temperature difference is important in hydronic boilers. Many of the older systems were designed using a 20 degree Delta T. This included the temperature rise through the boiler as well as the temperature drop from the supply to the return of the system. If the system Delta t is less than 20 degrees F, the system is not giving up its heat. If the delta T is higher, the flow may not be adequate. A wide temperature drop could cause Thermal Shock which can damage the boilers.

Condensing temperature - Operating a standard non condensing boiler below 140 degrees F will allow the flue gases to start to condense, which could destroy the boiler, flue and chimney. It might also void the warranty. Another effect is that soot may build on the fire side of the boiler, causing a very dangerous condition.

Connect Circulator to Fan Terminal - Most commercial thermostats allow continual fan operation during the occupied time. On smaller commercial buildings, we will sometimes control the pump using the fan terminal of the thermostat. During the occupied setting on the thermostat, the circulator will operate continuously. During the unoccupied time, the pump will only operate when there is a call for heat.

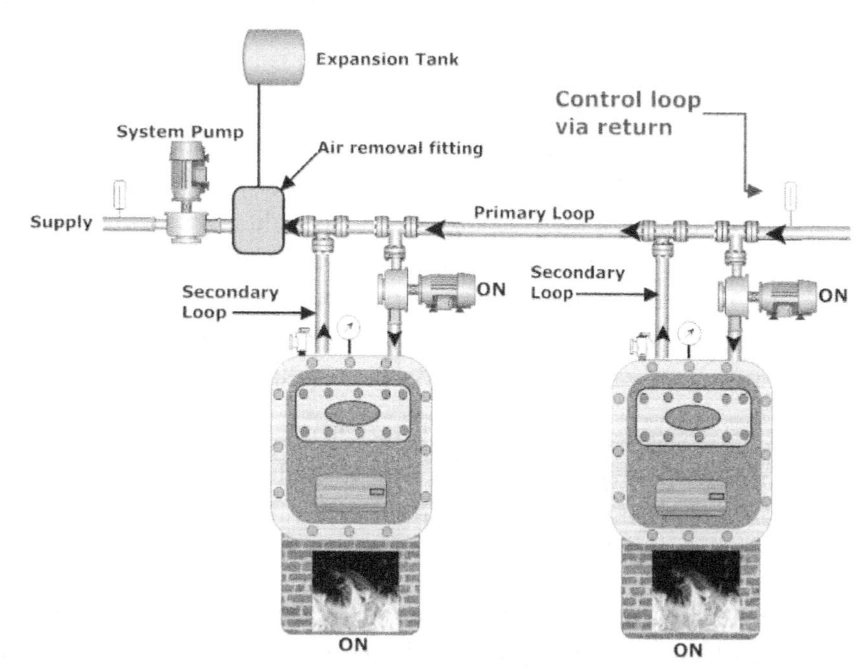

Sensor Location When controlling the heating loop, I have found that the best location for the sensor is on the return. If the sensor is on the supply, it will cause the boilers to short cycle.

Thermal purging - On small hydronic systems with large mass boilers, this has been a very cost effective solution. We will use either a two stage thermostat or a time delay relay. On a call for heat, the system circulator starts. If after a certain amount of time the thermostat is still calling for heat, the burner starts. This allows the system to use the residual heat in the boiler to heat the building. If there is not enough heat, the burner will start.

Bypass Valve A bypass valve is sometimes used to introduce warm supply water into the return loop to keep the boiler above the dew point or condensing temperature.

Zone Valves

Three Way Valve A three way valve could create thermal shock and damage the boiler as the cool water from the system meets the hot water inside the boiler, creating leaks if not installed correctly.

Zone Valves An apartment building that we serviced had a zone valve for every apartment. While this seemed to make sense to the installer, it did not work well. When several of the zone valves were closed, the flow through the open valves was excessive resulting in noisy operation. It sounded like the call of orca whales. Yes, I watch Animal Planet. This high velocity would not allow the water to give up its heat. The boiler short cycled and the apartments that were cold stayed cold. To remedy the problem, we had to disconnect the control wiring, manually open all the electric zone valves, and balance the flow. We balanced the flow by measuring the temperature drop from the supply to the return. We adjusted the flow for a twenty degree drop. We then installed a reset control and a remote thermostat in one of the apartments. In addition to being unable to control the building properly, the closed valves could damage the pump by dead heading the pump. When the zone valves were not open, there was insufficient flow through the boiler, damaging that. If I was servicing this building today, I would have installed a variable speed control on the circulator that would slow down when only a zone or two called for heat.

Radiator valves are commonly installed on radiators. Some of them are the self-contained ones that use thermal expansion to close the valve. The drawback to radiator valves is that it could cause the pump to dead head if they all shut off at the same time. It could also cause excessive flow through the open valves if the pump is operating at the same speed.

Dead Heading a pump is when all the flow is stopped for the pump by closing the valves. On the older pumps with positive displacement, this could have caused damage to the pump and mechanical seal. The caution with the smaller circulators today is that the water temperature could raise high enough from the pump impeller spinning to cause the water to flash to steam, ruining the pump.

When two way valves close, the system pressure may rise and could affect the pump. Pressure bypass is sometimes used with two way valves. The bypass control senses the pressure in the piping and will open a bypass valve to allow the flow back to the pump inlet. This reduces the chance of excessive velocity through the piping. Another way to eliminate the excessive flow is to use three way valves instead of two way valves as a way to control the velocity.

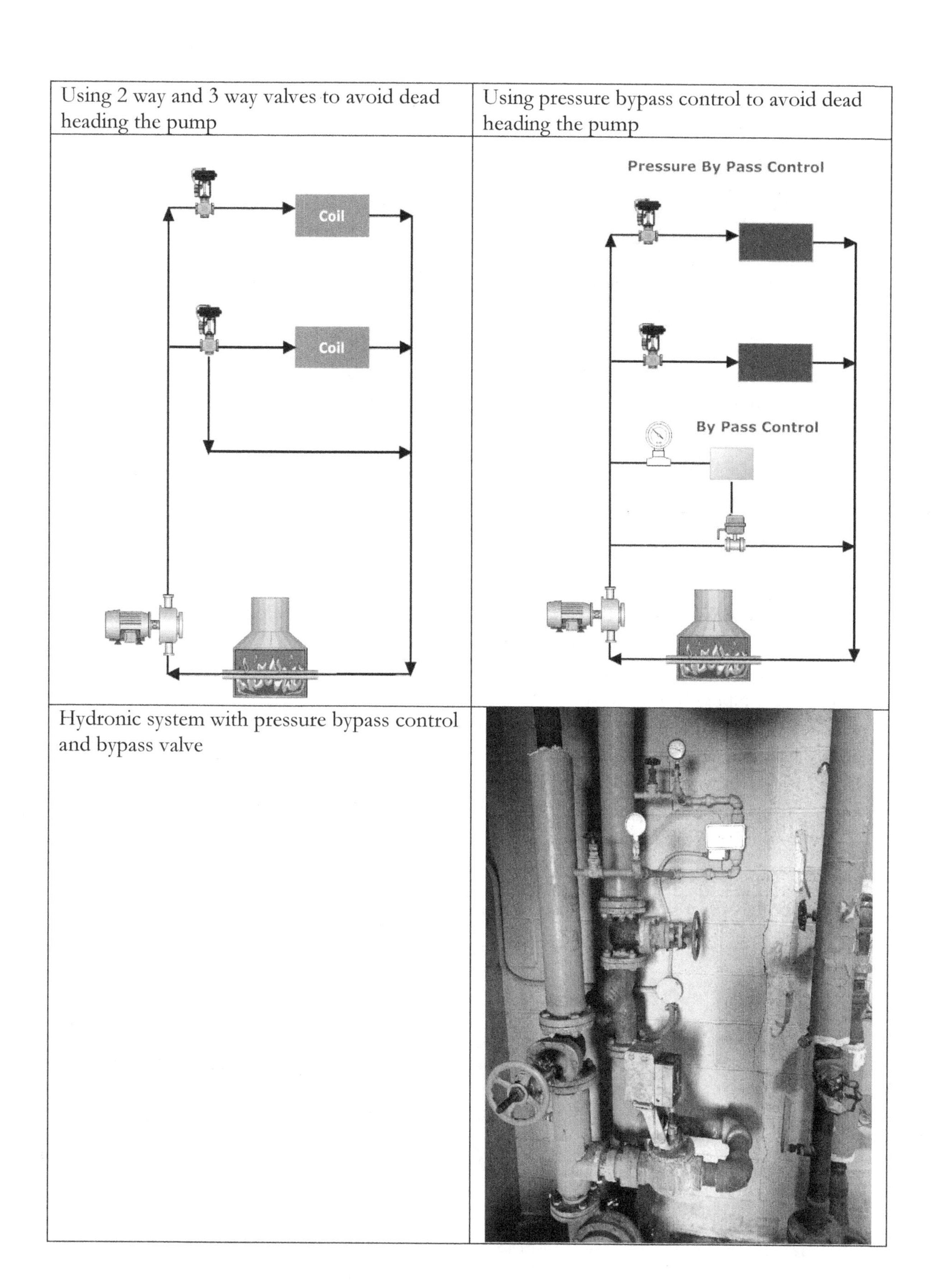

Using 2 way and 3 way valves to avoid dead heading the pump

Using pressure bypass control to avoid dead heading the pump

Hydronic system with pressure bypass control and bypass valve

Hydronic Low Water Cutoffs	
Float Style	**Probe Type**

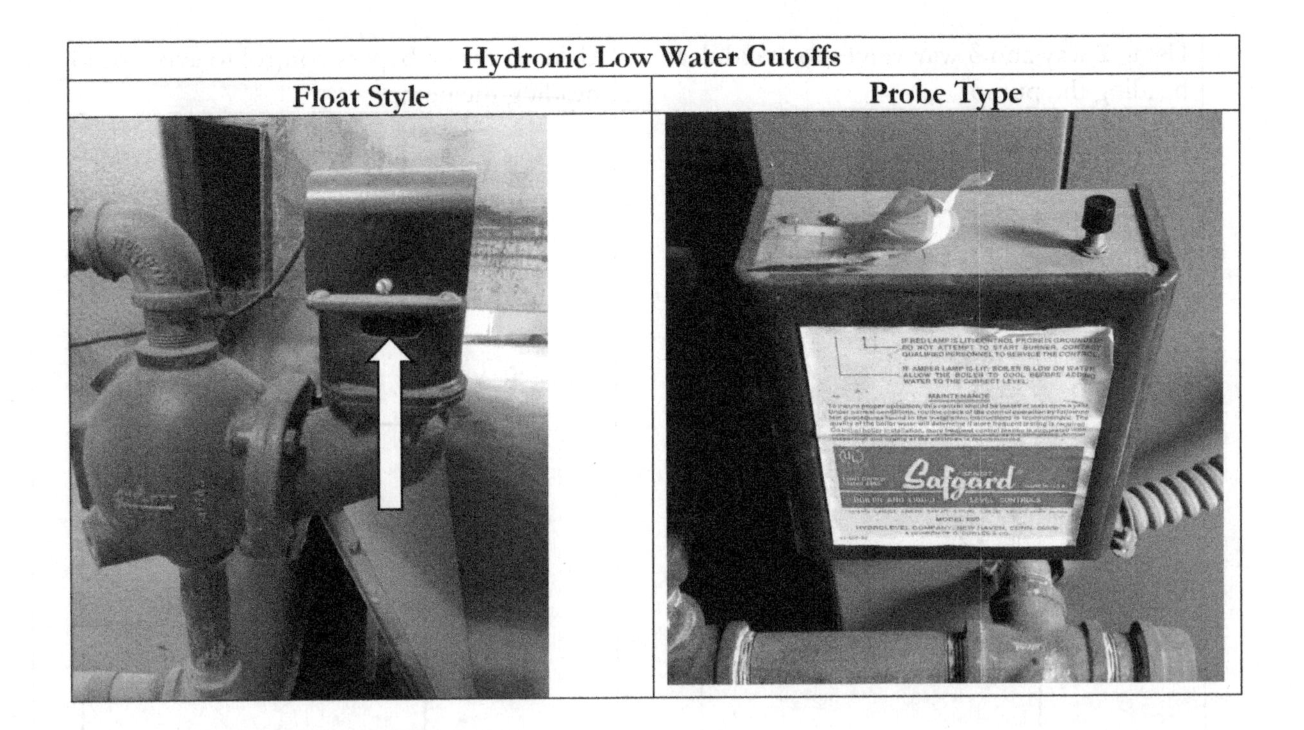

Hydronic Low Water Cutoff These low water cutoffs use either a float or probe to detect the water level. These are sometimes difficult to test as the system is filled with water and hard to drain the entire building to test the low water cutoff. Many hydronic systems have the low water cutoff in the piping above the boiler. If your locale adheres to the ASME CSD1 code, the low water cutoff should have a test button that allows you to test the control. On older boilers that do not have the test button, you have to be more innovative. These controls cannot be isolated from the system with valves. On a float type low water cutoff, you may be able to reach in with a narrow screwdriver to the linkages to verify that they are free and will shut off the boiler. Be careful not to bend the internal brass components.

Hydronic Low Water Cutoffs	
Some boilers use a flow switch instead of low water cutoff	The probe from a probe type low water cutoff

Probe Type Hydronic Low Water Cutoff These types of low water cutoffs use the conductivity of the water to verify the proper water level. The low water cutoff will send electrical voltage to the probe and if the probe is surrounded by water, the voltage will ground itself through the water and allow the burner to operate. If the low water cutoff does not have a test button, I will shut off the power to the boiler and remove the wire from the probe. I will then turn on the burner to see if it will start. Be careful as there will be live electricity on the wire. To check the probe, verify that the probe is grounded to the boiler when there is water in the boiler. Excessive Teflon tape on the threads will sometimes impede the connection with the boiler. Hydrolevel suggests that you remove and clean the probe on their low water cutoffs once a year on commercial boilers and once every five years on residential boilers. The probe is in direct contact with the water so you will need to drain the boiler below the probe elevation to remove it or you will have a face full of hot water. Scale sometimes forms on the probe and could affect the conductivity. When cleaning the probe, consult with the manufacturer as Hydrolevel suggests using a scouring pad while McDonnell & Miller suggests a soft cloth. It is common for air pockets to form around the probe so you have to slowly unscrew the probe and be sure not to completely unscrew it and allow the air to vent.

Flow switch Some boilers, like water tube tubes, use a flow switch instead of a low water cutoff. This should be treated with the same respect as the low water cutoff and inspected regularly. I like to pull the flap out and check the condition. If the flap is worn or fragile, I will replace it. I was called to a jobsite where the customer said the boiler sounded like "A Cat in Heat" When we arrived, we found that the boiler was scaled and the client had "jumped out" the flow switch. The boiler had to be descaled and the flow switch replaced.

Hydronic Boiler System Pressure How much pressure should be on the pressure temperature gauge on a hydronic boiler? It depends on the height of the building above the boiler. We need to get the hot water to the highest radiator or heat emitter. Many think that the pump or circulator is used to reach those radiators far away. In reality, it is system pressure, which is shown on the Pressure Temperature Altitude or PTA gauge on the boiler, to reach the radiators. That job falls to the water feeder or PRV, pressure reducing valve. How much pressure do you need? Well, one pound of pressure will raise water 2.3 feet. A good rule of thumb for tall buildings is that the boiler pressure should be

half the height of the highest radiator. If the highest radiator is 50 feet high, the system fill pressure should be 25 pounds.

Why are some boilers in the penthouse? As you know, one pound of pressure will raise water 2.3 feet. Conversely, a column of water has a weight of that same amount. If you have a tall building, the weight of the water at the bottom could be substantial. Consider a twenty story high rise building and the highest radiator is at the top floor 300 feet above the boiler, the weight of water would be around 130 pounds at the bottom of the building. Some very tall buildings have mechanical rooms at different floors because of the pressure exerted from the column of water.

Water Feeder This is the feeder that fills the system with water. The bottom of the feed water valve has an internal screen that should be cleaned as part of the boiler service.	

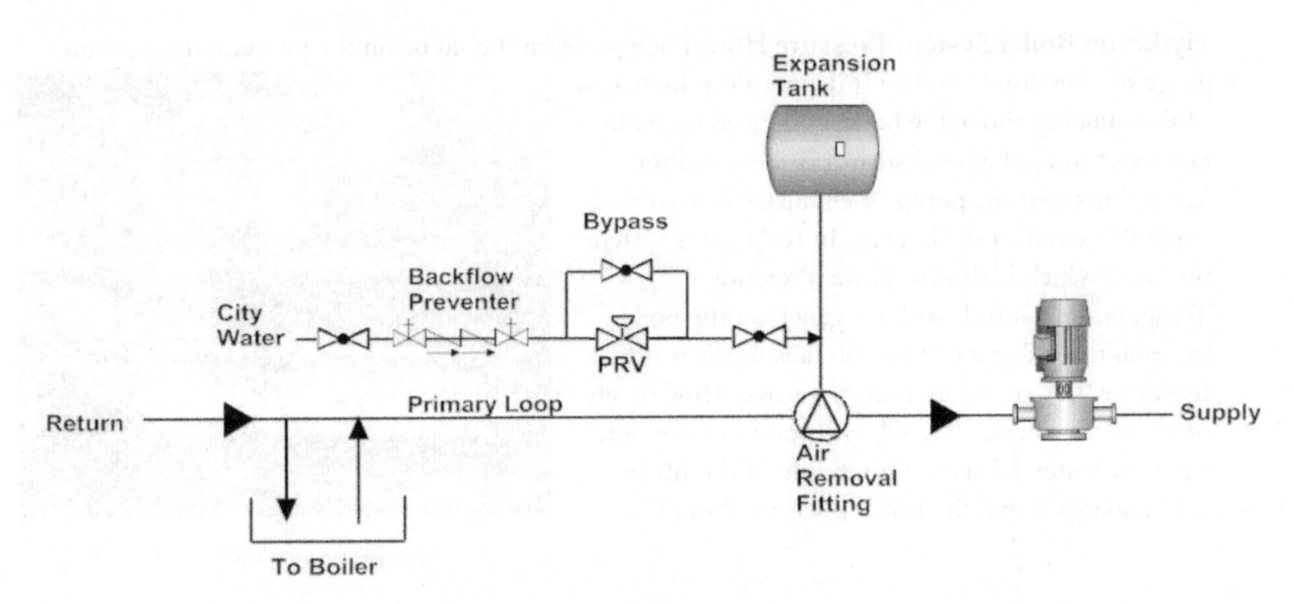

The residential feed valves are usually adjustable from 10 - 25 psi. The pressure reducing valve is usually set at 12 pounds of pressure at the factory. This pressure was used to assure that water would reach the top radiator of a traditional two story home. On commercial systems, it may require more pressure to get it to the highest radiator. One pound of pressure will raise water 2.3 feet. If the highest radiator is on the top floor of a five story building, it may be 50 feet high. To see how much pressure we need for that elevation, divide the height of the highest radiator by 2.3 feet. In this example, it would be 50 feet divided by 2.3 psi = 21.7 psi or round it up to 22 psi. That just assures that we have the water to that elevation, Bell and Gossett suggests that you add an additional 3-4 psi to allow the air to vent and the water to not flash to steam inside the pipe. That will take us to between 25 - 26 psi. That may be at the far end of our water feeder rating and will require a larger one.

Feed valve piping is important in a hydronic system. A bypass loop around the feeder should be installed to allow quick fill of the system. Make sure that the bypass piping does not also bypass the backflow preventer. A boiler inspector made us re-pipe our bypass loop as it bypassed the backflow and the feed valve. He felt that the bypass could be opened and the dirty boiler water could be fed back to the building's potable water supply.

Maintenance The water feeder could have an internal strainer that requires cleaning. Watts recommends cleaning the screen on theirs twice per year. I was able to resurrect an inoperable water feeder by cleaning the strainer and the scale buildup inside it.

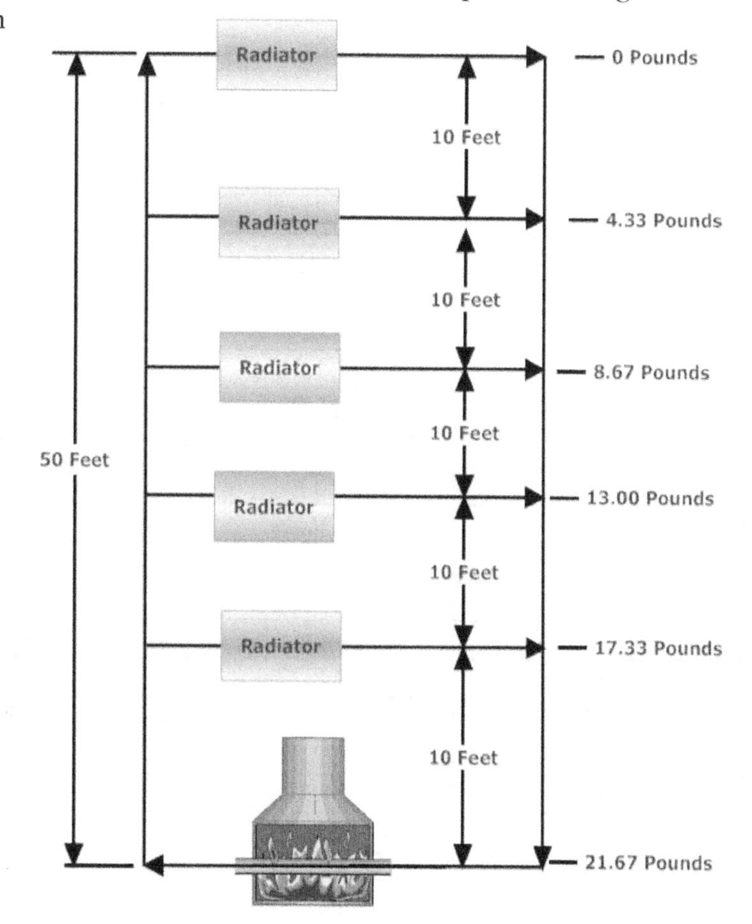

Open or Closed? There is debate in the industry as to whether to leave the water feeder open to the system or to close the valve to isolate it. I would say that you should follow the manufacturer's recommendations. For instance, the installation manual for a Watts Series 1156F feeder says, *This valve must always be kept open when the system is in operation.* If the system contains glycol, I would be hesitant to leave the valve open as the makeup water could dilute the glycol concentration.

Backflow Preventer Most municipalities require a backflow preventer to be installed on the feed water line to the building. You should also install one on the water feed pipe to the boilers. This prevents that nasty black water inside the boiler, filled with chemicals to be back fed into the potable (drinking) water for the building which could be dangerous if ingested. On hydronic system, it will be installed on the feed water connection to the system, usually in the pipe that goes from the boiler to the expansion tank. On steam boilers, they should be on the pipe that feeds fresh water to the system, usually on the boiler feed unit. Some of the old boilers had makeup water fed directly into the boiler, which I do not recommend. If so, the backflow preventer should be installed on the makeup water pipe to the boiler. According to section 608.16.2 of the International Plumbing Code, backflow preventers are required where the potable water feeds the boiler.

Hot Water System Makeup: Minimum connection size shall be 10% of largest system pipe or 1", whichever is greater 20" pipe should equal a 2" connection.

The following chart shows how much pressure correlates with height in feet of water.

Water Pressure to Feet Head

Pounds Per Sq Inch	Feet Head	Pounds Per Sq Inch	Feet Head
1	2.31	**100**	230.90
2	4.62	**110**	253.98
3	6.93	**120**	277.07
4	9.24	**130**	300.16
5	11.54	**140**	323.25
6	13.85	**150**	346.34
7	16.16	**160**	369.43
8	18.47	**170**	392.52
9	20.78	**180**	415.61
10	23.09	**200**	461.78
15	34.63	**250**	577.24
20	46.18	**300**	692.69
25	57.72	**350**	808.13
30	69.27	**400**	922.58
40	92.36	**500**	1,154.48
50	115.45	**600**	1,385.39
60	138.54	**700**	1,616.3
70	161.63	**800**	1,847.2
80	184.72	**900**	2,078.1
90	207.81	**1,000**	2,309.00

System Pressure required to reach highest hydronic radiator		
Radiator Height	Pressure required to reach radiator	Suggested Pressure Setting*
10	4.3	8.3
20	8.6	12.6
30	12.9	16.9
40	17.2	21.2
50	21.5	25.5
60	25.8	29.8
70	30.1	34.1
80	34.4	38.4
90	38.7	42.7
100	43	47
150	64.5	68.5
200	86	90
250	107.5	111.5
300	129	133
350	150.5	154.5
400	172	176
450	193.5	197.5
***Includes the extra 4 pounds that Bell and Gossett suggests.**		

Rules of Thumb to Estimate System Volume In some instances, you may need to estimate the water volume in the system. This is used for estimating water treatment, adding glycol, or simply amazing friends at a cocktail party. The following are some rules of thumb to estimate the system volume.

- Multiply steel compression tank volume by 5.
- 35 – 50 gallons per Boiler HP
- Pump GPM x 4
- Compression Tank volume is 20% of system volume
- Rated tonnage of system x 10 gallons

PTA Gauge The gauge on hydronic boilers is called the Pressure Temperature Altitude or PTA gauge. The gauge usually has two needles with three readings. The gauge will display the following:

Pressure of the hydronic system at the boiler

Temperature of water inside the boiler

Altitude will tell you how high the water can be in the building. This is directly tied to the system pressure.

Hydronic Piping Sizes Pipe sizing is important in a hydronic system. Under sizing the piping is far more detrimental to the system than over sizing the piping. Undersized piping could lead to

- Boiler short cycling
- Reduced seasonal efficiency
- Comfort complaints
- Increased heating costs
- Noisy operation

Oversized piping increases the installation costs.

Normal Water Temperature Ever since December 1899, most non-condensing hydronic systems were designed to supply 180^0 F at the outdoor design temperature of the locale with a temperature drop or ΔT(Delta T) of 20^0 degrees F. On conventional, non-condensing boilers, operation below 140^0 degrees F on the return or 160^0 degrees F on the supply could result in the condensation of flue gases. Flue gas condensation will destroy the boiler and flue.

The condensing boilers are designed to operate at much lower temperatures than a traditional boiler. Condensing boilers do not actually condense until the water temperature is below 140^0 F. If a condensing boiler is operated above 140^0F, it has efficiencies in the mid to upper 80% range.

Bleeding the Hydronic System Bleeding a hydronic system could take several hours on a large building. Sometimes, you may have trapped air pockets in the hydronic system. I have had success freeing air bound parts of the system by starting and stopping the pumps and by increasing the system pressure. I have also been able to close and open zone valves to try forcing more water into the air bound zone.

Do I have an Air Problem? Dan Holohan, boiler expert from HeatingHelp.Com, says that if you are not getting air from a radiator when you bleed it, you do not have an air problem.

Circulator According to industry averages, based mounted pumps will last 20 years and pipe mounted pumps will last 10 years. The pumps should be maintained on a regular basis. On the small circulators that are sealed, there is not much maintenance. I will feel the pump as it runs for excessive vibration. Be careful as it may be hot to the touch. I will also check the amperage of the pump motor to see if it is drawing the proper amps. If it is drawing excessive amps, the motor windings may be wearing. If the amps are low, there could be a problem with the impeller. The amp draw is an indication of the work that the motor is doing.

Pump Pressure Gauges If the pump has pressure gauges, this can tell you how many gallons per minute the pump is moving if you have the pump curve.

Listen to the motor for sounds like rubbing or growling. Worn bearings may sound like two pieces of sandpaper rubbing, especially during starting and stopping.

Continuous Flow? It is common for the pumps on commercial systems to operate continuously all winter. On large systems, variable frequency drives should be considered. For example, a 2 horsepower pump will cost the building owner about $6.46 per day to operate. This would be based upon a kwh cost of $0.1346. If you estimate a heating season of October 1 to March 31, the owner would pay $1,169.26 for the electrical consumption of the pump for the season. A variable speed drive could cut that cost dramatically. When using a variable speed drive, be sure that you have the proper flow through the boiler as inadequate flow could damage the boiler.

Boiler Flow Verify that there is adequate flow when the boiler fires. A rule of thumb is that the boiler should have at least three GPM per boiler horsepower. This should be verified with the boiler manufacturer.

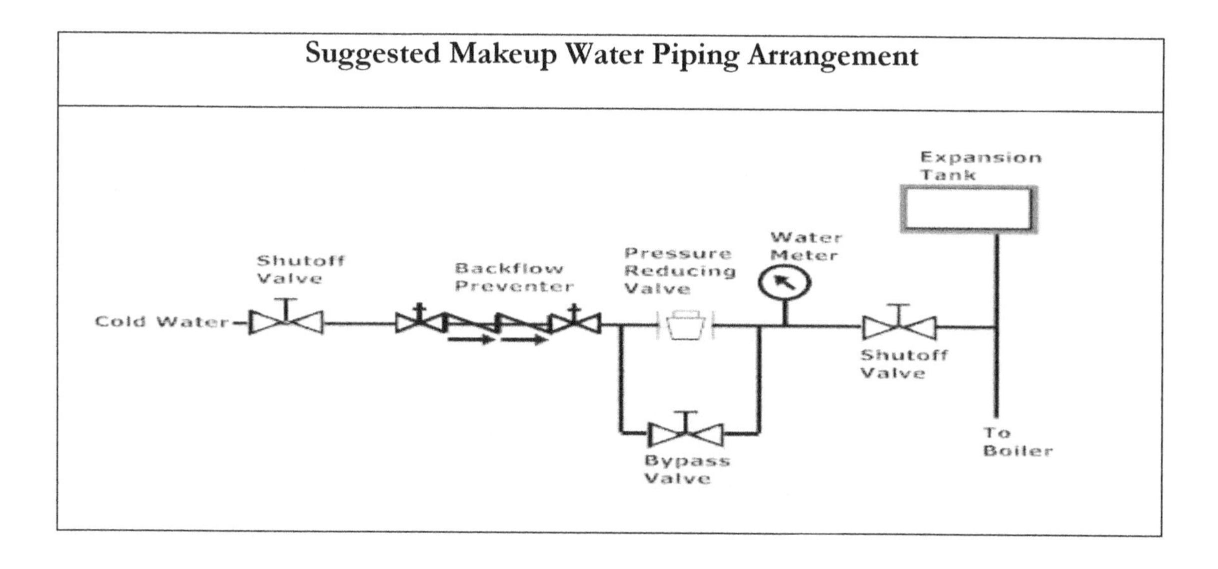

Suggested Makeup Water Piping Arrangement

Boiler Design Temperature Most hydronic systems were designed to provide 180^{0} F water at the "Outdoor Design Temperature". The "Outdoor Design Temperature" was originally the 1% design temperature. This meant that in an average winter, the outdoor temperature would be colder than the design temperature 1% of the time. Now, the industry has adopted the 2 1/2% temperature. This means that in an average winter, the temperature will be at or below the design temperature 2 1/2% of the time. For example, if we have 4,000 heating hours during a typical heating season and our design temperature is 0^{0}F, 120 hours will be below 0^{0}F. The outdoor design temperature is different in almost every locale. For instance, the heating design temperature for Fairbanks, Alaska is -47^{0}F while the design temperature for Jacksonville, FL is 32^{0}F.

Velocity The velocity in a hydronic system should range from 2 to 4 ½ feet per second in occupied areas. If the flow is excessive, the system will be noisy. The velocity can be slightly higher in unoccupied areas. Flows greater than 6 feet per second can erode copper. Excessive velocity can also lower the heating capacity of the system. Flows below 2 feet per second will not capture the air in the system, leading to air bound parts of the heating system. In case you were wondering about velocity:

Feet per Second	Miles per Hour
1	0.68
2	1.36
3	2.05
4	2.73
5	3.41
6	4.09
7	4.77
8	5.45
9	6.14
10	6.82
Shaded areas are velocities that are outside the recommended ranges for hydronic systems.	

Reset Control Assuming that our original heating system was designed for a heat loss of 1,000,000 Btuh at the design outdoor temperature of 0^0 F with a hot water supply temperature of 180^0 F. The system was most likely also designed for a 20-degree Delta T or temperature drop. That means that the return water temperature will be twenty degrees lower than the supply temperature. Our design was also based on a certain indoor temperature. In this example, we will use 72^0 F. The greater the temperature difference between the outside air temperature and the inside temperature, the greater or faster the building loses its heat to the outside.

As the outside temperature rises, the difference between the inside temperature and the outside temperature decreases. Since the temperature difference is less, our heat loss rate declines. If we used 180^0 F to successfully heat our building at 0^0 F, it makes sense that we could heat our building at 30^0F with a lower water temperature. A reset control will assume that as well. A reset control will reduce the system temperature as the outside temperature rises. A common reset ratio in the industry was one to one. That means that the system water temperature drops one degree for every degree that the outside air rises above the design temperature. One of the problems with this reset schedule was that the boiler flue gases would condense if the water temperature dropped to below 140^0F. In reality, we could only reset the water by 20^0 F without fear of condensation. Verify with the boiler manufacturer how low the boilers can operate.

Typical One to One Reset Schedule	
Outside Temperature	Supply Temperature F
0^0 Design temperature	180^0 F
20^0	160^0 F
40^0	140^0 F*
60^0	120^0 F*

*Standard boiler could be condensing at this temperature.

Copper Pipe Maximum Hydronic Flow Rates

Based on 20 degree F Delta T

Pipe Size	Maximum Flow GPM	Btuh
½"	1 1/2	15,000
¾"	4	40,000
1"	8	80,000
1 ¼"	14	140,000
1 ½"	22	220,000
2"	45	450,000
2 ½"	85	850,000
3"	130	1,300,000

Hydronic Pipe Sizes

PEX Piping Maximum Hydronic Flow Rates

Pipe Size	**3/8"**	½"	**5/8"**	¾"	**1"**
Max GPM	1.2	2	4	6	9.5
BTUH	12,000	20,000	40,000	60,000	95,000

Steel Pipe Maximum Hydronic Flow Rates

Based on 20 degree F Delta T

Pipe Size	Maximum Flow GPM	Btuh
½"	2	15,000
¾"	4	40,000
1"	8	80,000
1 ¼"	16	140,000
1 ½"	25	220,000
2"	50	450,000
2 ½"	80	850,000
3"	140	1,300,000
4"	300	3,000,000
5'	550	5,500,000
6"	850	8,500,000
8	1,800	18,000,000
10"	3,200	32,000,000
12"	5,000	50,000,000

Circulator Information

Modern hydronic systems use a pump or circulator to distribute heat from the boiler to the building. Some of the old systems relied on gravity to move the hot water around the system. It was a rather resourceful system as the hot water was lighter and would rise to the radiators while the cold water was heavier and would drop. To do this, the pipes were oversized. It is rare to see a gravity system still in operation but we see many systems that have been converted to a pump circulator. One of the issues with the converted systems is they have a tendency to overheat the first floor and under heat the top floor. Some of the older gravity systems had an orifice in the radiator to the top floor that will require relocation to the bottom floor.

Multiple Zone Circulators

Multiple pumps on the system return were a common way of providing zone control as each zone pump would have a thermostat that would control the pump. While it did provide better temperature control, it could adversely affect the heating system in a couple ways The first is that you are not pumping away from the compression tank so the inlet of the pump could actually go sub atmospheric or into a vacuum, pulling air into the system through air vents and valve packing. This air will attack the metal surfaces inside the heating system causing leaks. The other drawback is that unless all the pumps are operating, there may be insufficient flow through the boiler which could damage the boiler. Many manufacturers have a minimum flow that they require for the boilers. Inadequate flow could cause thermal shock.

Pump Bearing Assembly The picture shows the bearing assembly on a small circulator. Each heating season, the bearing assembly requires a few drops of oil to allow long life.	

Pump Couplings Some circulators use a separate motor and pump and use a pump coupling to connect the two. Other components on a hydronic system may include a **triple duty valve**. This valve is a combination of check valve, balance valve and shutoff valve. Another common accessory for a circulator is a **suction diffuser**. It is on the inlet to the pump. It has a strainer to limit the dirt that may flow into the pump. It also has straightening vanes to reduce noise and the stress on the impeller due to uneven flow.

Isolation Valves could cause some serious damage if they are closed when the pump is running. This is called Dead Heading the pump. If the valves are closed, the temperature and pressure inside the pump housing rises to a dangerous level very quickly and could damage the pump or could even cause it to explode. Consider this: If a small B& G Series 100 pump with only a 1/12 HP motor is running with closed valves, the temperature inside the circulator will rise 50^{0}F per hour.

Expansion joint Some pumps require an expansion joint to absorb the vibration inherent in a moving pump. Be careful that the pump anchors actually allow the expansion joint to capture the vibration.

Pump pressure gauges Most pumps will have pressure gauges on the inlet and discharge of the pump for diagnosing and servicing the pump.

Pump Mounting Options According to ASHRAE, a base mounted pump has a life expectancy of 20 years while a pipe mounted pump will last only 10. The following show the different mounting arrangements.

Pump Impeller The water enters through the center of the impeller and exits out the sides.	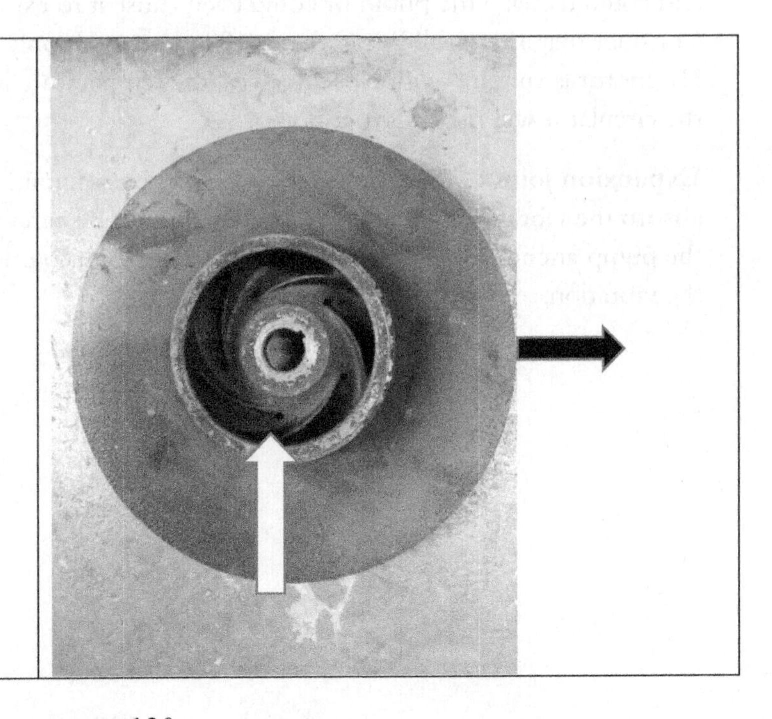

Base Mounted Pump

Pipe Mounted Pump

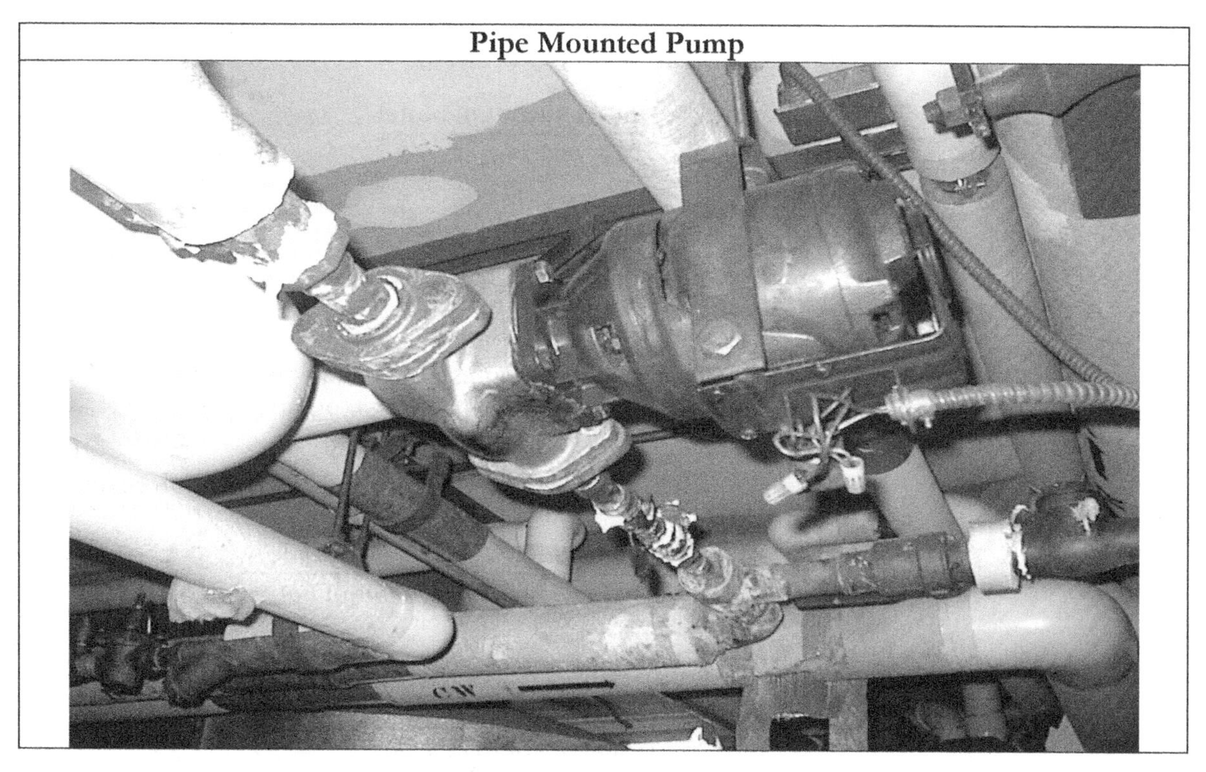

Sizing a Circulator

There are a couple short cuts to sizing a pump for a boiler. Most boilers are designed for a 20 to 30 degree rise through the boiler. To size a pump for a boiler and maintain a 20 degree rise through the boiler, divide the output of the boiler by 10,000 to get the proper GPM for a 20 degree rise. If the boiler can handle a 30 degree rise, divide the output of the boiler by 15,000 to get the proper GPM or Gallons per Minute.
Example: To see if the existing 40 GPM pump on a project is large enough for the boiler, let us look at the equipment. Our new boiler has a rated output of 800,000 with a design temperature rise of 20 degrees F. The existing pump is 40 GPM. If we divide the boiler output by 10,000, we see that the boiler will require an 80 GPM pump. This is double the GPM of the existing pump. Our flow would be half and the temperature rise would be double, possibly ruining the new boiler. If the boiler can handle a 30 degree rise, we could divide it by 15,000 and find that the boiler will require a 53 GPM pump. The existing pump is still too small for this boiler.

If you would like to see how I arrived at the 10,000 or 15,000 number, the following is the formula:

$$\text{GPM} = \frac{\text{Rated output of boiler}}{8.33 * 60 * \triangle\ °\text{F}}$$

or

$$\text{GPM} = \frac{\text{Rated output of boiler}}{500 * \triangle\ °\text{F}}$$

500 = 8.33 * 60

GPM = Gallons per minute flow rate
8.33 = Weight of a gallon of water
60 = Converts the formula from hours to minutes aka Gallons per Minute GPM
△ °F Temperature rise through boiler is usually about 20-30 degrees F.

The following is the actual formula for a 20 degree rise for the 800,000 Btuh boiler:

$$\text{GPM} = \frac{800{,}000\ Btuh\ (Output\ of\ boiler)}{8.33 * 60 * Temperature\ rise\ through\ boiler}$$

$$\text{GPM} = \frac{800{,}000\ Btuh\ (Output\ of\ boiler)}{500 * 20\ Degree\ rise}$$

500 = 8.33 * 60

$$80 \text{ GPM} = \frac{800{,}000 \; Btuh \; (Output \; of \; boiler)}{10{,}000}$$

The following is the actual formula for a 30 degree rise for the 800,000 Btuh boiler:

$$\text{GPM} = \frac{800{,}000 \; Btuh \; (Output \; of \; boiler)}{8.33 * 60 * Temperature \; rise \; through \; boiler}$$

$$\text{GPM} = \frac{800{,}000 \; Btuh \; (Output \; of \; boiler)}{500 * 30 \; Degree \; rise}$$

$$53 \text{ GPM} = \frac{800{,}000 \; Btuh \; (Output \; of \; boiler)}{15{,}000}$$

The following is a chart to help you verify the pump size based upon common delta T

Boiler Output	*Temp rise through boiler*			Boiler Output	*Temp rise through boiler*	
	20°F GPM	*30°F GPM*			*20°F GPM*	*30°F GPM*
500,0000	50	33		**1,500,000**	150	100
600,000	60	40		**1,750,000**	175	117
700,000	70	47		**2,000,000**	200	133
800,000	80	53		**2,500,000**	250	167
900,000	90	60		**3,000,000**	300	200
1,000,000	100	67		**4,000,000**	400	267
1,250,000	125	83				

Calculate pump head

1 Measure longest run in feet. Include both supply and return.

2 Multiply by 1.5 to calculate fittings and valves

3 Multiply by 0.04 (4' head for each 100' of pipe ensures quiet operation)

For example, 1,000 feet is longest run

$$1{,}000 \times 1.5 \times .04 = 60 \text{ feet of head}$$

or

Measure longest run, divide by 100 feet, and multiply by 6 feet to get pump head.

$$1000 \div 100 = 10$$

$$10 \times 6 = 60 \text{ feet}$$

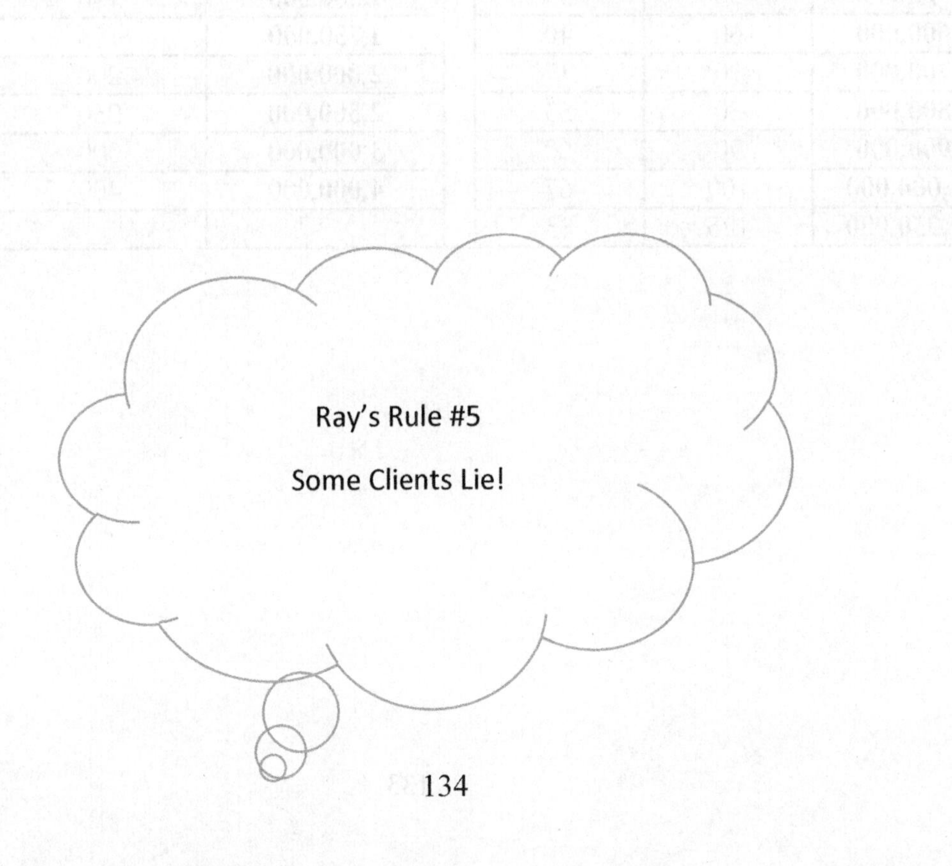

Pumping away The location of the circulator to the system is important to the proper operation of the system. Many boilers have the circulator on the return piping and while the pump will still circulate water on the return, there are several disadvantages to that location.

Why are the circulators on the return? There are two schools of thought on this question. The first is that the seals on the older pumps could not handle the higher temperatures on the supply piping so they were installed on the return piping. The second consideration is that the pump will be less susceptible to damage if the pump is on the side rather than the top of the boiler. The drawback to having the pump on the return is that it could allow air into the system. The pump creates flow by creating a pressure differential in the piping. The discharge of the pump will be at a higher pressure than the inlet. If both pressures are the same, the system will have no flow. So, if we have the pump on the return, the discharge of the pump will be facing the boiler and the compression tank. When the pump turns on, the positive pump discharge pressure is negated by the compression tank as it is absorbed by the compression tank. The pump does not realize that so it will try to create a pressure differential in any way that it can. If it cannot increase the discharge pressure, it will increase the suction. In some applications, this can actually pull the inlet negative and air will enter the piping. This could cause corrosion inside the boilers and pipes. It will also create comfort complaints as the air will find its way to the most inconvenient locations and you will have air bound radiators that have to be bled regularly. In severe applications, you may have to bleed the radiators several times a year.

The preferred location for the pump is on the discharge piping just past the takeoff of the compression tank. This is referred to as "Pumping Away." When the pump starts it will increase the discharge pressure on the outlet and the inlet stays at the pressure setting of the water feeder. In this way, the system will be less susceptible to internal corrosion and air bound heat emitters.

Why does my compression tank flood? In some instances, the pump may actually pull the air from the old metal compression tanks and introduce it into the piping system if the piping is incorrect.

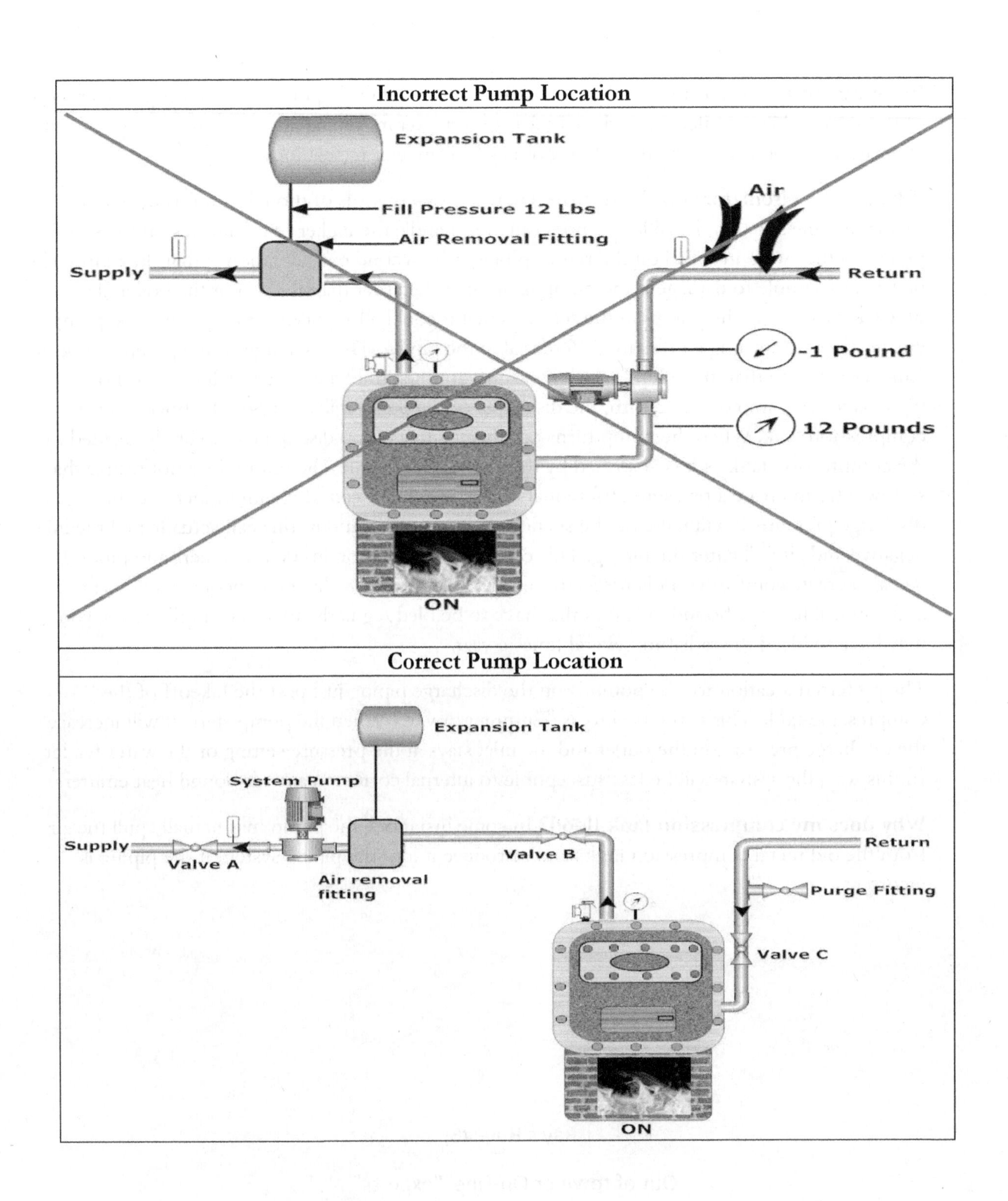
Incorrect Pump Location
Expansion Tank
Fill Pressure 12 Lbs
Air Removal Fitting
Air
Supply
Return
-1 Pound
12 Pounds
ON
Correct Pump Location
Expansion Tank
System Pump
Supply
Valve A
Air removal fitting
Valve B
Return
Purge Fitting
Valve C
ON

Overheating Idle Zone? When you have two or more loops each with its own circulator, you could experience ghost flow through the idle zones. For example, if loop B is running, you could have ghost flow backwards through Loop A. This could lead to overheating the idle zone that Loop A handles. To eliminate the chance of that happening, flow control valves are often used. A flow control valve is like a weighted check valve. It is only opened when the zone pump for that zone is operating.

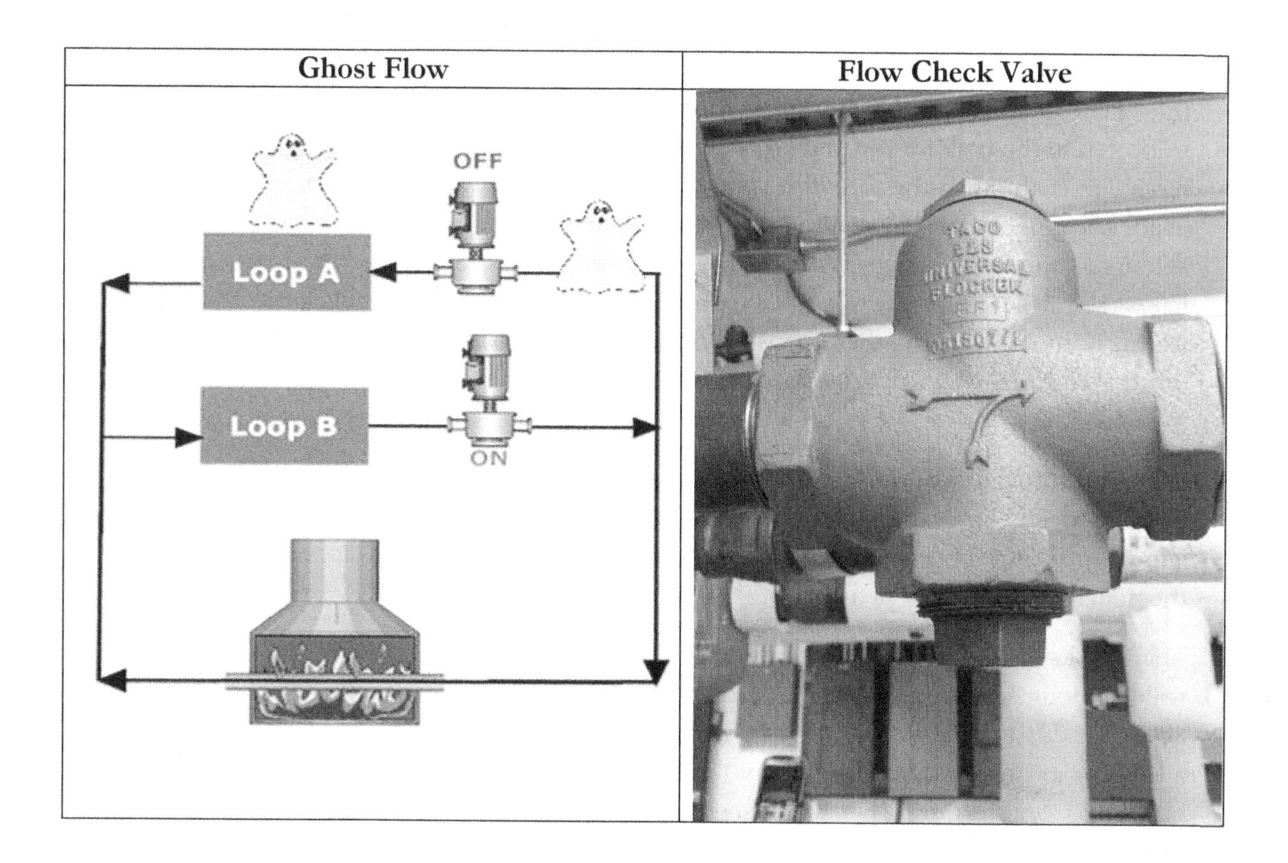

The first heating system in the White House was installed in 1809

Compression or Expansion Tanks When water is heated from 65⁰F to 180⁰F, it will expand at 3% of its volume. This expansion of the water has to be controlled or managed. This is typically done with a compression tank. Although they are frequently referred to as either a compression or expansion tank, there is a difference. The expansion tank was used on old gravity hydronic systems. It was located in the highest part of the building and was open to atmosphere. When water was heated, it expanded and the level rose in the open tank. The compression tank is located in the boiler room and is a cylinder which has an air cushion on the top of the tank. When water is heated, the expansion compresses the air inside the tank. When the water cools and the system pressure reduces, the air cushion is greater. When looking at the gauge glass on the compression tank, it should be about one quarter to halfway up when the water is cool and about 3/4 of the way up the gauge glass when the water is warm. Without this tank, the expansion could cause the water to leak through the valve packing or lift the boiler relief valve. You may see the relief valve discharge piping weeping when the tank is flooded.

Servicing the Tank If the water level is difficult to see, hold a pencil behind the gauge glass. The pencil will appear broken if the gauge is full of water. If it is empty, the pencil will appear to be intact.

Many tanks have an Airtrol fitting that has a long tube that will go about 2/3 of the way up the tank. This allows the air to be added to the top of the tank. Perhaps you played a game as a youth, holding your thumb over the top and pulling it from the drink. The liquid stayed in the straw until you removed your thumb. That is what happens with the compression tank when the Airtrol fitting is not working

If the tank is flooded, look for the following:

Tank Leaks The tank should be inspected for leaks. Look at the top of the tank for pin holes as this will allow the tank to flood.

Gauge Glass One of the common issues with flooding tanks is that the washers and nuts on top of the gauge glass are leaking air. Verify they are tight and not leaking. They will dry out after several years. I always keep some extra rubber gauge glass washers and brass rings to replace them if needed.

System Pressure Verify that the system pressure is correct. If the pressure is too high, it could cause the tank to flood.

A way to test if the compression tank is flooded or the piping is plugged to the tank is to watch the boiler PTA or Pressure Temperature Altitude gauge as the boiler fires. If the pressure climbs when the water heats, this usually means that the tank is flooded or the piping to the tank is plugged.

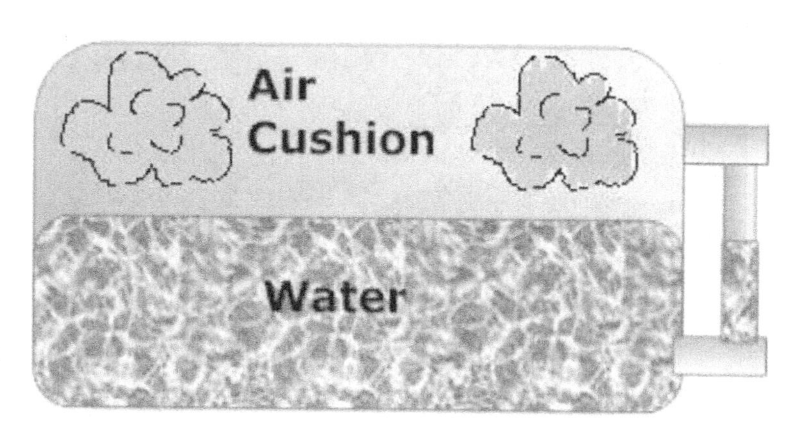

If the compression tank has an internal diaphragm, verify its integrity and the internal pressure is correct. The air pressure is checked with a fitting similar to one on your vehicle tire. As a matter of fact, you use a tire pressure gauge to check the pressure. It is like an inner tube in a tire. The air pressure should be checked while it is empty and not connected to the piping.

The following is the expansion in inches of different piping:

Thermal Expansion of Piping Material in inches per 100 feet from 32 Deg F

Temperature Deg F	Carbon & Carbon Moly Steel	Cast Iron	Copper
32	0	0	0
100	0.5	0.5	0.8
150	0.8	0.8	1.4
200	1.2	1.2	2.0
250	1.7	1.5	2.7
300	2.0	1.9	3.3
350	2.5	2.3	4.0
400	2.9	2.7	4.7
500	3.8	3.5	6.0
600	4.8	4.4	7.4
700	5.9	5.3	9.0

Horizontal Compression Tank	Vertical Compression Tank

Small compression tank

Rule of Thumb for Compression Tank Sizing

Based upon the following: Entering Pressure 10 pounds, Maximum Pressure 25 pounds
Entering Temperature 40°F, Maximum Temperature 220°F

Steel Piping					
System Capacity in Gallons	Closed Compression Tank	Diaphragm Tank	System Capacity in Gallons	Closed Compression Tank	Diaphragm Tank
200	39	23	**1,700**	327	195
300	58	34	**1,800**	347	206
400	77	46	**1,900**	366	218
500	96	57	**2,000**	385	229
600	116	69	**2,500**	481	287
700	135	80	**3,000**	578	344
800	154	92	**3,500**	674	401
900	173	103	**4,000**	770	458
1,000	193	115	**4,500**	867	516
1,100	212	126	**5,000**	963	573
1,200	231	138	**6,000**	1,156	688
1,300	250	149	**7,000**	1,348	802
1,400	270	160	**8,000**	1,541	917
1,500	289	172	**9,000**	1,733	1,032
1,600	308	183	**10,000**	1,926	1,146

Copper Piping					
System Capacity in Gallons	Closed Compression Tank	Diaphragm Tank	System Capacity in Gallons	Closed Compression Tank	Diaphragm Tank
200	37	22	**1,700**	315	188
300	56	33	**1,800**	334	199
400	74	44	**1,900**	352	210
500	93	55	**2,000**	371	221
600	111	66	**2,500**	463	276
700	130	77	**3,000**	556	331
800	148	88	**3,500**	649	386
900	167	99	**4,000**	742	441
1,000	185	110	**4,500**	834	496
1,100	204	121	**5,000**	927	552
1,200	222	132	**6,000**	1,112	662
1,300	241	143	**7,000**	1,298	772
1,400	260	154	**8,000**	1,483	883
1,500	278	165	**9,000**	1,668	993
1,600	297	177	**10,000**	1,854	1,103

Multiple Compression Tanks Rules of thumb

- The vertical pipe between air separator and tanks should be 3/4" or larger.
- The horizontal pipe to the manifold tanks should be as follows:
- 3/4" if pipe run is less than 7 feet long.
- 1" if pipe run is 7 feet to 20 feet long.
- 1 1/4" if pipe run is 21-40 feet long.
- 1 1/2" if pipe run is between 40-100 feet long.

Tank Manifold Pipe Sizing

- 1" for 2 tanks.
- 1 1/4" for 3-4 tanks.
- 1 1/2" for 5 or more tanks.

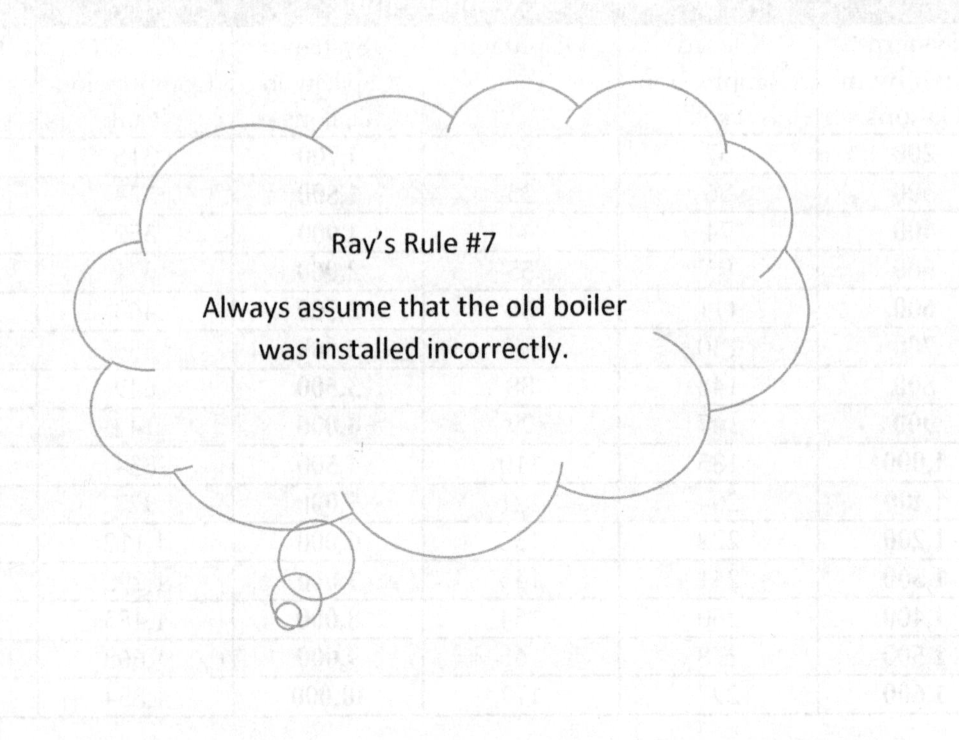

Air Removal Air is the enemy of any hydronic system. It causes internal corrosion of the pipes and boilers. It could also impede the flow of water to areas of the building resulting in comfort complaints. To eliminate air from the system, an air removal fitting is usually used. This device will capture the air and either send it to the compression tank or vent it to the boiler room. The air removal fitting should be installed after boilers and before the system pump. The pipe between the compression tank and the boiler is usually where the makeup water connection for the system is piped. Air removal fittings work best when a straight piece of pipe 18" or longer precedes the fitting.

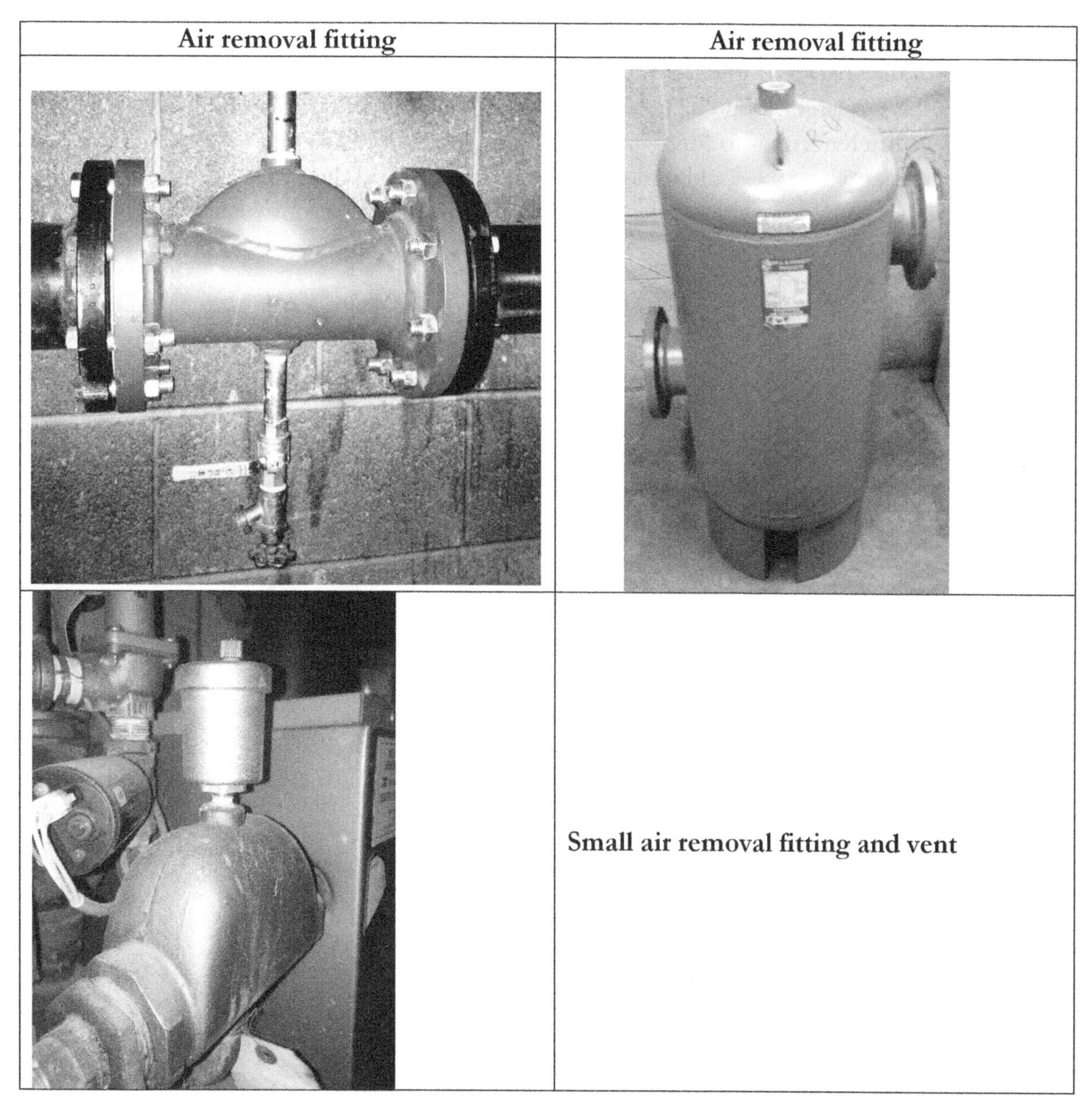

Air removal fitting	Air removal fitting
	Small air removal fitting and vent

With any hydronic system, air must be removed before hot water could enter the terminal equipment or piping. Check the existing piping or terminal equipment to see how the air could be removed. See whether there are vents in the piping or equipment. Some have automatic vents while most have manual vents. Many of the manual vents use a special key called the radiator vent key that is available in most heating supply houses and big box stores. I like to give the client a few extra keys so they could remove the air from the system themselves.

Getting rid of the stubborn air pocket. A certain wing in a building was having a difficult time heating. When we opened the vents, the system spilled water from them. The maintenance department had repaired a leaking coil and could not get heat to another zone after the repair. We correctly diagnosed that the system had an air blockage to the zone somewhere in the system. The challenge was to get the air blockage moved and removed. To do so, we had to perform a series of tasks that finally worked. We would cycle the pumps on and off. This allowed the natural buoyancy of the air to raise to the top of the system. We also raised the system pressure to try moving this large air bubble inside the piping to an area where it could be removed. We would close different zones to force more flow through the blocked ones. It finally moved the air pocket to a place where we could remove it. Another way to rid the system of air is to install about an ounce of Dawn dish detergent. This breaks the air bubbles into smaller units and they can be removed easier.

Did you know that water will
expand at 1,600 times its volume
when converted to steam?

Purge fitting 1 On smaller systems, it is a good idea to install a purge fitting in the boiler room on the return piping. It is simply a hose bib that allows the air to go through the system and be vented in the boiler room. Although this does not eliminate all the air, it does rid the system of a great majority. Try to imagine how long it would take to vent a large building of the air using an 1/8" air vent in the top. For this to work, you close valves B and C. Connect hose to purge fitting. Open valve A and allow water to run through the hose until all the air bubbles are gone. You could then open valves B and C. To get rid of air inside the boiler, you may have to open the relief valve.

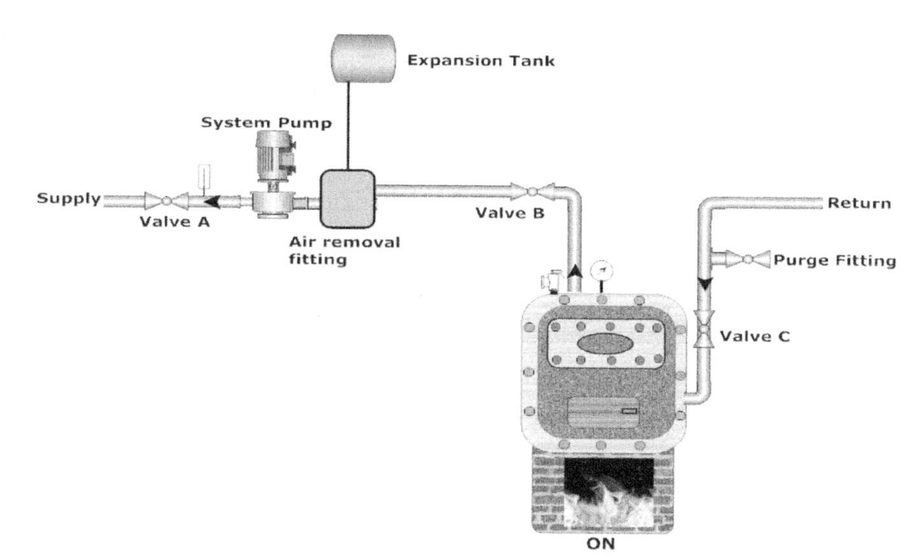

Air Purge Part 2 This is another variation on the purge fitting. It will allow you to vent most of the air from your system without having air vents installed in the piping. Here is how it works: Close Valve B. Open Valve C and Valve A. Connect hose to boiler drain 1 on the run of the tee. Open boiler drain 1. Fill your system through the fill bypass. Put the hose into a bucket. Run water until the bubbles stop coming through the water in the bucket.

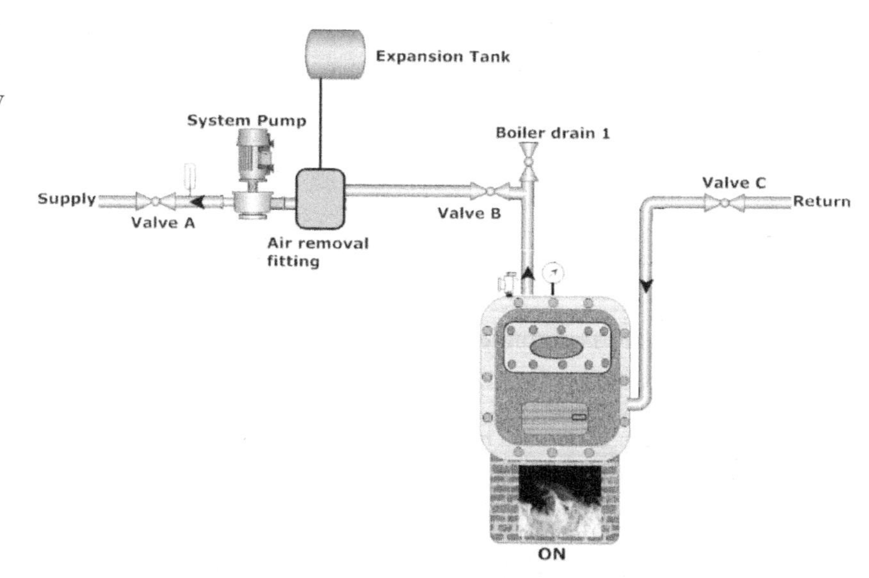

Servicing a Condensing Boiler

When servicing a condensing boiler, there are substantial differences than servicing the traditional type hydronic boilers.

Maintenance The maintenance is higher on condensing boilers than the standard hydronic boilers due to the condensing flue gases. If you follow the manufacturer's recommendations, there may be major disassembly that is recommended yearly including the removal of the burner and accessing the fireside of the boiler for cleaning.

Combustion Air Since most condensing boilers have combustion air directly vented from the outside, the dirt could find its way into the burner and more frequent cleanings are required.

Gaskets Prior to opening the boiler for the tear down, you should purchase a set of gaskets from the boiler manufacturer. I have found that the gaskets are not necessarily a stock item. Many manufacturers have a maintenance kit that contains all the gaskets and other materials required for the yearly service. It may be a good idea to order them and store them at the customer's location.

Burner Many of the condensing boilers use a mesh type burner. These have to be carefully cleaned and inspected.

pH The pH level on condensing boilers is critical. They are much more prone to damage from the pH levels in the water. They pH levels should be checked regularly and verified with the manufacturer.

Water Treatment The water treatment for condensing boilers is more critical on condensing boilers. Standard water treatment may damage the boiler. Consult with the manufacturer about their requirements.

Glycol If using glycol inside the boiler, the type should be verified with the manufacturer as some glycol can destroy the boiler.

Proprietary Parts The condensing boilers have proprietary parts and it is a good idea to stock the flame rod or igniter for burner in your truck.

Controls These boilers also have different looking controls than you see on traditional boilers. Many of the controls may be part of the main control board. It is a good idea to have the maintenance manual with you when servicing the boilers.

Condensate Drain Since the combustion air is coming from outside, dirt becomes entrained in the air and will sometimes result in plugged drains for the condensate. The drain should be checked monthly.

Operating Temperature The condensing boiler is designed to operate at much lower temperatures than standard hydronic boilers. If you operate the boiler at the traditional 180 degrees F that most systems were designed for, the condensing boiler will not condense. It will have efficiencies in the mid to upper 80%. The colder the water, the more efficient the boiler is.

Low Pressure Steam

Servicing a low pressure steam system is completely different than service on a hydronic system. You have to look at steam systems more holistically as each component will affect other parts of the system. The boiler guru, Dan Holohan, always says that the problem and the solution are rarely in the same room. For example, we had a client with a portion of one zone of their building not heating. Each zone had an automatic valve and a thermostat in the zone. We checked the zone valve and found that it was opening but only a small portion of the area would heat. We thought that perhaps the return piping was plugged. After several agonizing hours, we removed the vacuum breaker that was installed on the outlet of the zone valve in the boiler room three stories below the problem room. As soon as we did, the building had heat everywhere. We had to replace the vacuum breaker.

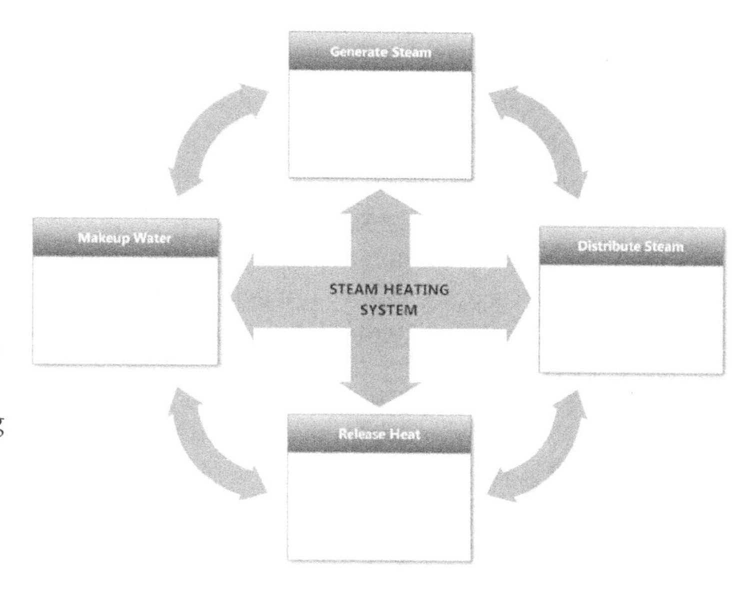

Can Air Escape? When working on steam systems, the piping will either have steam, condensate or air. For steam to enter a pipe or radiator, the air has to be removed. Steam and air cannot take up the same space because they are both gases. When trouble shooting steam systems, ask yourself this simple question, If I were air, how would I escape the pipe? If the system uses traps, the air will leave with the condensate via the traps. On the drawings below, the black arrow simulates air in the system. Look on the outlet piping of the trap for sagging pipes as this will create a water seal and stop the air.

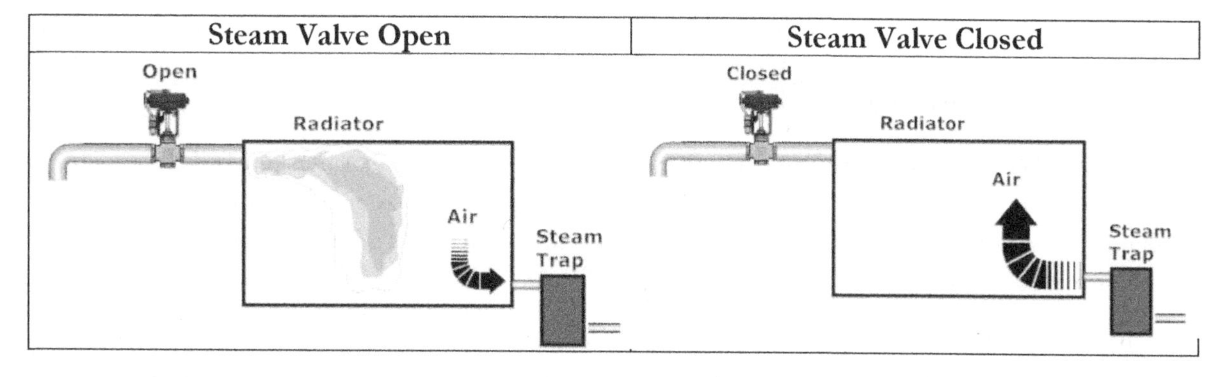

Vacuum is not your friend If the steam system has a vacuum forming in the piping, it will drive you crazy. A deep vacuum can pull the water out of the boiler and even the boiler feed tank. A slight vacuum can cause many an ulcer for the service technician. That is why I like using a compound gauge or a digital manometer that will show me if a vacuum is forming. How does a vacuum form in a steam system? When water changes to seam, it will expand at 1,600 times its volume. When steam condenses, that volume suddenly collapses. If the air cannot get back in to take the place of the steam, a vacuum will form.

Vacuum in Steam Piping We had a project where the owner installed thermostatically controlled radiator valves on every radiator in the building to save money. When the boiler shut off and the steam began to condense, the steam system developed a deep vacuum in the piping. On some occasions, it would actually pull the condensate from the boiler feed unit and flood the boiler. Once we installed a vacuum breaker, the problems disappeared. You can test for a vacuum forming by using a compound gauge or digital manometer. In the old days, we would blow smoke from a cigarette and see if it gets sucked in. You may be able to hear the system sucking if you open a fitting. I will either loosen a union or open the strainer to see if it sucks air.

Steam Pressure Ever since December 1899, most low pressure steam systems were designed to operate on 2 pounds of steam pressure or less. If the system is set higher than that, you should investigate the reason. If the client has a steam to water heat exchanger, many were designed using higher steam pressure, usually 5-7 Psig.

Banging at the Synagogue We were called to look at a heating issue at a local synagogue and met with the rabbi. During the meeting, I asked what sort of problems they were experiencing and he mentioned high fuel costs, uneven heating and banging. When a steam system is banging, it is important to note when during the heating cycle that the banging occurs as it will aid in diagnosing the problem. The banging is also called "Water Hammer." When I asked Rabbi Weiss when the banging started, he said, "Only when I am about to speak." I asked if it could be one of the attendees and he laughed. It turns out the banging started when the boiler started heating. The asbestos insulation was removed and never replaced and there were sags in the piping where water would lay. When the steam hit the cold water, it would start banging. If the water hammer happens later in the heat cycle, it could be caused by improper system piping or plugged return pipes.

Radiator Not Hot All The Way Across? This is a common question asked and the radiator will only heat the whole way on the coldest days. The radiator was sized to heat the room to 70 degrees F on the coldest day using the entire heating surface. On days when it is warmer, the radiator will heat the room with less heating surface.

Boiler Staging When using modular steam boilers, you cannot control them like you would a hydronic system. I prefer staging all the boilers on at the same time and once steam pressure is built up, you could start staging the boilers off. It is analogous to an airplane taking off. The pilot gives full thrust to the engines to get off the runway. Once the plane hits cruising altitude, the pilot pulls back the throttle and allows the plane to coast. That is how a steam system with multiple boilers works best.

Radiator valves are commonly installed on radiators. Some of them are the self-contained ones that use thermal expansion to close the valve. If these type valves are used on steam systems, it could cause a vacuum to form inside the system. If these are used, a vacuum breaker should be installed. Some automatic radiator valves have built in vacuum breakers.

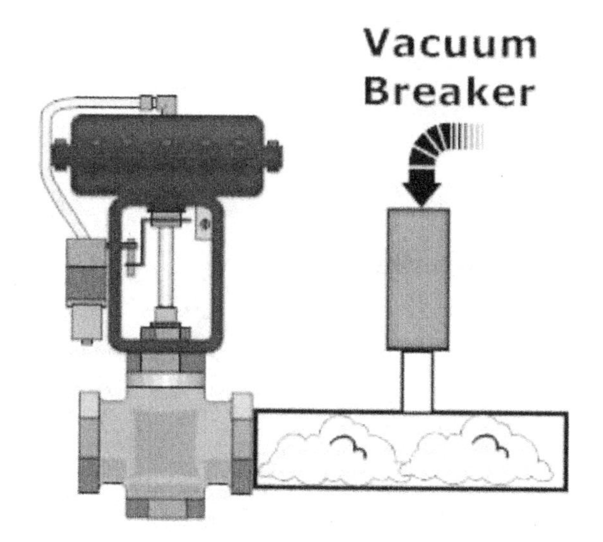

Zone Valves When these are used on steam systems, they could also cause a vacuum that could pull the water from the boiler. A vacuum breaker should be used when zone valves are installed.

Did you know that the worst boiler accident ever was when the SS Sultana exploded and killed over 1,800 people?

Steam Low Water Cutoffs	
Combination Pump Control Low Water Cutoff	Auxiliary Low Water Cutoff

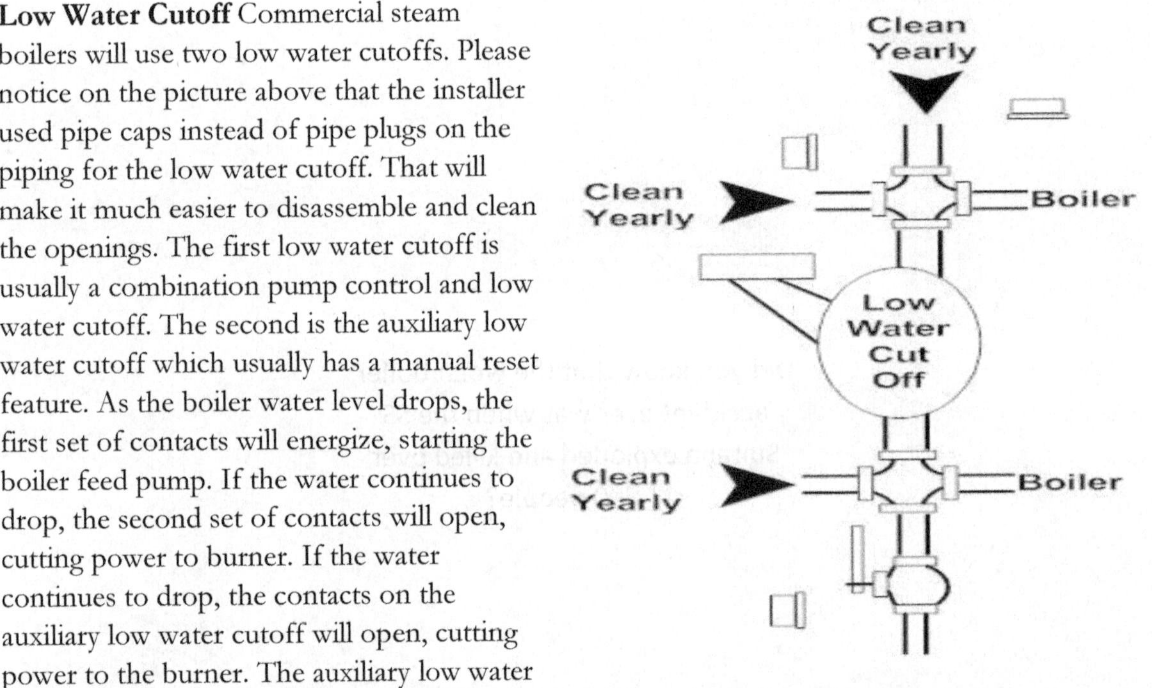

Low Water Cutoff Commercial steam boilers will use two low water cutoffs. Please notice on the picture above that the installer used pipe caps instead of pipe plugs on the piping for the low water cutoff. That will make it much easier to disassemble and clean the openings. The first low water cutoff is usually a combination pump control and low water cutoff. The second is the auxiliary low water cutoff which usually has a manual reset feature. As the boiler water level drops, the first set of contacts will energize, starting the boiler feed pump. If the water continues to drop, the second set of contacts will open, cutting power to burner. If the water continues to drop, the contacts on the auxiliary low water cutoff will open, cutting power to the burner. The auxiliary low water cutoff should be a manual reset control that will need to be reset before the power to the burner is restored. The reason for a manual reset on the low water cutoff is to alert you that the primary low water cutoff did not operate correctly. The auxiliary low water cutoff is installed at a lower elevation than the primary low water cutoff. Some low water cutoffs have an audible alarm that will notify the owner of a low water condition.

Be careful connecting the boiler feed pump to the combination pump control/ low water cutoff contacts as the contacts may not be able to handle the amperage of the pump motor. In most instances, you will need to use an additional relay rated for the motor amperage.

Servicing the Low Water Cutoff

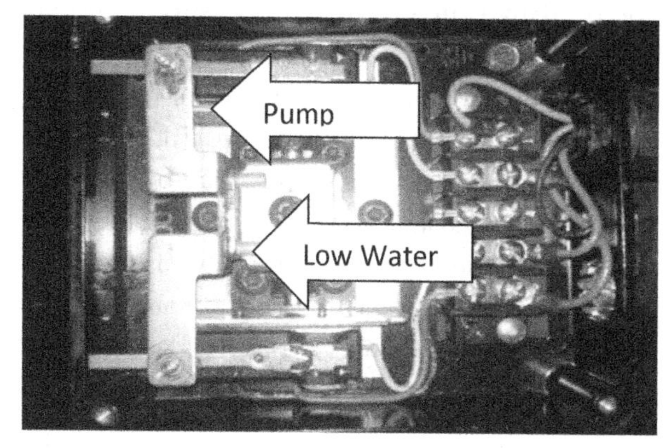

The picture shows the inside of a McDonnell Miller #150 combination low water cutoff and pump control. One internal float will start or stop the boiler feed pump. The other will shut off the burner if the water level drops too low. You could watch the water level inside the gauge glass.

The low water cutoff should be serviced on a regular basis. When servicing these, you should purchase a gasket that will be used to re-attach the head assembly. Typically, you will disassemble these yearly. The inside of the bowl will require cleaning so that the float moves freely inside the chamber. The new low water cutoffs use a probe instead. When reassembling the head of the low water cutoff, both sides have to be cleaned and free of grit or the gasket will not seal. The piping connections to the boiler should be inspected and cleaned.

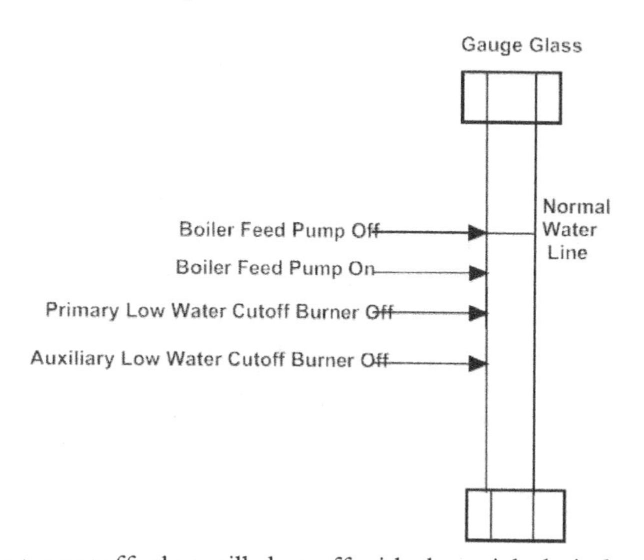

When I start up the boiler, I will mark the normal water levels on the gauge glass. This makes it easier when trouble shooting in the future.

Simulate actual conditions I like to simulate the actual field conditions when testing the low water cutoffs. On commercial buildings with a boiler feed or condensate tank, I will shut off the pump and allow the boiler to steam the water out of the boiler until it shuts off. When you quickly open the blowdown valve, that is not simulating the actual conditions. I have found that some low water cutoffs that will shut off with the quick drain but will keep the burner running with a slow drain test due to dirt or scale formation inside the float chamber. On smaller residential boilers, I will slowly drain the boiler by opening the bottom drain valve. Be careful so that it will shut off when the test is complete. I keep hose caps in my truck in case the valves do not shut off completely.

Maintenance The McDonnell & Miller #150 low water cutoff / pump control is probably the most common commercial primary low water cutoff control that I see in the field and the following are some of the maintenance recommendations:

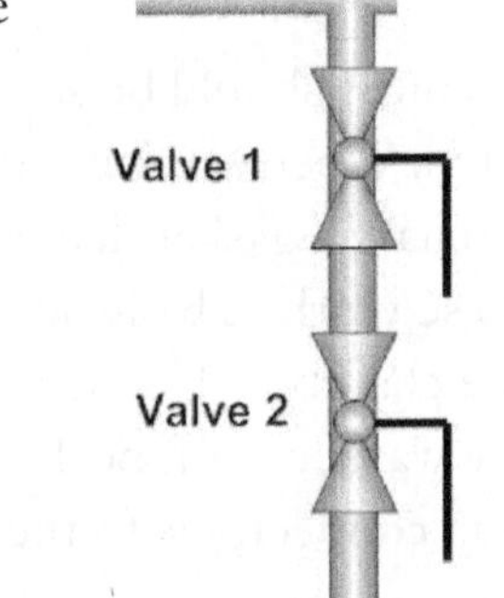

Blowing down the control at pressure. A potential hazard when testing the operation of a low water cutoff when the boiler is under pressure is the danger of the hot steam hitting the cool stagnant water in the control. It could cause water hammer which is very dangerous. One of the ways to avoid that danger is to install two valves in the blow down pipe. To avoid water hammer, open upper valve 1 and then slowly open lower valve 2. Once you verify that the control operates properly, slowly close valve 2 and then close valve 1.

Blow Down Blow down the control once a week for low pressure steam boilers and daily if the boiler is operating above 15 psig.

Remove head assembly and check water side components yearly. Replace components that are worn, corroded, damaged.

Inspect the float assembly and equalizing piping annually. Remove debris and sediment.

McDonnell Millers recommends replacing the head assembly every five years.

If the area contains hard water or the system has dirty water, you may need to perform the maintenance more often. A client of ours has extremely hard water and required maintenance frequently so we installed a water softener on the boiler makeup water and the maintenance dropped dramatically. Water softeners on steam systems are a good idea as they allow you to reduce the scale inside the boiler.

Many boiler inspectors will require that you open the low water cutoff so the inside bowl can be inspected for scale or build up. Some clients will ask you to perform this task while others will do it in-house.

Plugged Low Water Cutoff The picture to the right shows a low water cutoff that was plugged with mud and dirt. The boiler and low water cutoff were only two years old and the boiler suffered a major catastrophic failure as a result of the buildup inside the control. The mud inside the control kept the float from falling so the boiler continued to fire without water, which is never a good thing. Luckily, there were no injuries or damage beside the boiler.

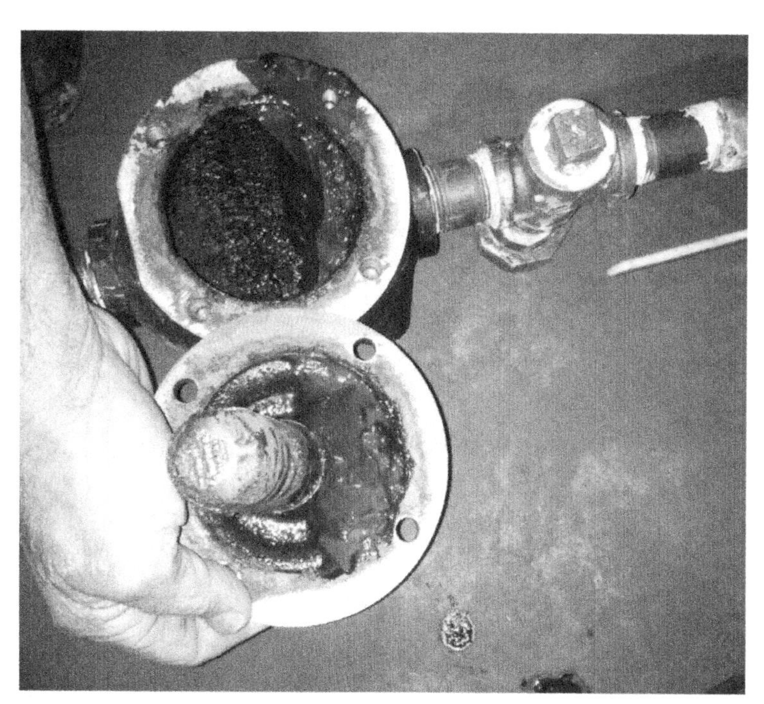

Probe Type Steam Low Water Cutoff These types of low water cutoffs use the conductivity of the water to verify the proper water level. The low water cutoff will send electrical voltage to the probe and if the probe is surrounded by water, the voltage will ground itself through the water and allow the burner to operate. If the low water cutoff does not have a test button, I will shut off the power to the boiler and remove the wire from the probe. I will then turn on the burner to see if it will start. Be careful as there will be electricity on the wire. To check the probe, verify that the probe is grounded to the boiler when there is water in the boiler. Excessive Teflon tape on the threads will sometimes impede the connection with the boiler. Hydrolevel suggests that you remove and clean the probe on their low water cutoffs once a year on commercial boilers and once every five years on residential boilers. The probe is in direct contact with the water so you will need to drain the boiler below the probe elevation to remove it or you will have a face full of hot water. Scale sometimes forms on the probe and could affect the conductivity. When cleaning the probe, consult with the manufacturer as Hydrolevel suggests using a scouring pad while McDonnell & Miller suggests a soft cloth.

Auxiliary Low Water Cutoff Commercial low pressure steam boilers require an additional low water cutoff that is installed just below the primary low water cutoff and just above the lowest permissible water level of the boiler. This control usually is a manual reset control which means that you have to press the reset button once the water level is established. The reason for using the manual reset is to tell you that the primary low water cutoff did not work or shut off the boiler. I use the slow drain test to check this as well. This should be opened and the float chamber inspected and cleaned yearly as well. If you have bouncing water levels inside the boiler, that may trip the auxiliary low water cutoff. Watch the characteristics of the water level inside the boiler gauge glass.

Probe Type Low Water Cutoff	Probe Type Low Water Cutoff
Courtesy: Warrick Controls	

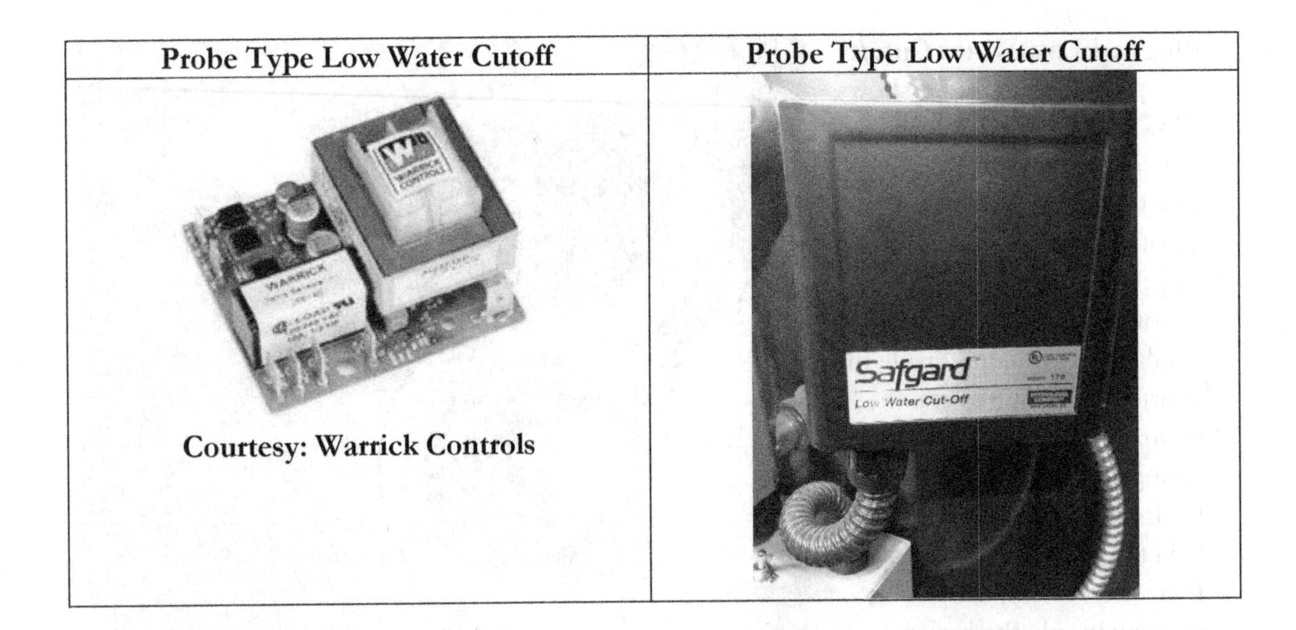

Water Feeder Many of the older boilers used a water feeder like the one in the picture. This may be dangerous as it allows cold water to enter a hot boiler. Most new steam boilers want the makeup water fed directly to the boiler feed tank so that it gets warmed by the returning condensate and does not shock the boiler. ASME CSD1* CW-120 forbids water being made up directly to boiler. *ASME CSD1 codes pertain to boilers that are between 400,000 and 12,500,000 Btuh.

Steam Pressure Controls

Older Pressuretrol	Newer Pressuretrol
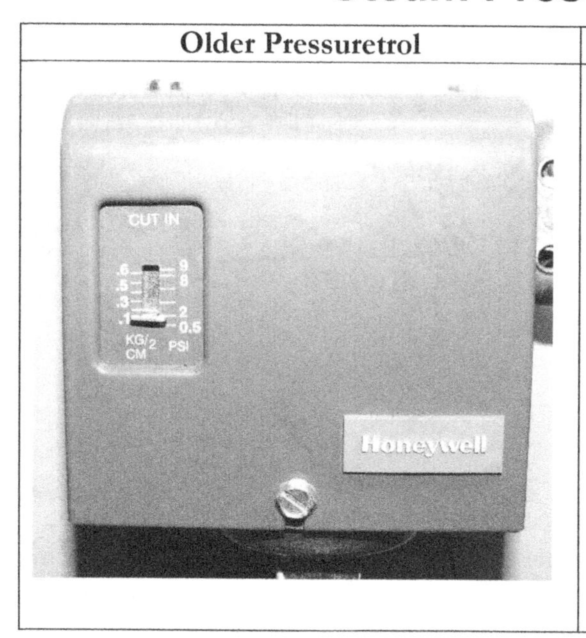	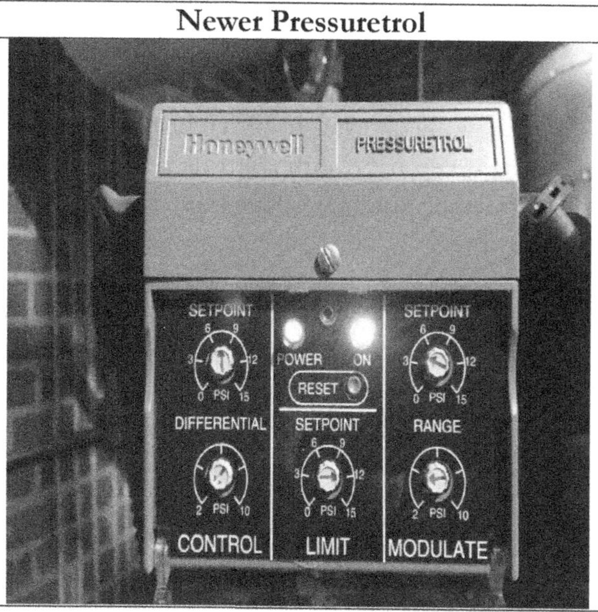

Normal Steam Operating Pressure In December 1899, the industry adopted 2 psi as the standard operating pressure for comfort low pressure steam systems. If the current boiler is set at a higher pressure than that, investigate why. It could cost the owner increased fuel costs.

How Does Steam Heat Work? When heating water, it roughly takes One Btu to raise one pound of water one degree F. That is, until the water temperature reaches 212^{0}F at atmospheric pressure. The boiler keeps heating that same pound of water trying to make steam. It requires about 970 additional Btus to get that pound of water to change to steam. This is called Latent Heat. The steam leaves the boiler at about 40 miles per hour looking for something cold. When it finds a cold radiator, it will release that 970 Btus and revert back to water or condensate. The real heating of the building occurs when that latent heat is surrendered into a space.

Mercury pressure control with siphon

Some thoughts about the old steam systems

When the old steam systems used coal as their fuel source, they were designed to run all day long. The boiler tender would keep the fire going inside the boiler. It was rare in the winter that the pipes were allowed to cool. When using a new boiler, the pipes cool in between the firing sequences. This could lead to carry over and bouncing water levels.

Installation When installing the controls and pressure gauge on a steam boiler, I prefer attaching the controls to different fittings on the boiler. If you have both controls coming from one opening and that opening plugs, you have lost control of your boiler. If you use two different pipe fittings, the likelihood of both being plugged is greatly lowered. ***Never install a shutoff valve in the piping from the boiler to the control.***

Many newer steam boilers use their own water seal with larger pipes that allow it to be cleaned easier. When installing a water seal for the controls, I would suggest using pipes and caps rather than pipe plugs as they are easier to remove. The pipes for the pressure controls and gauges should be disassembled and cleaned yearly to be sure they are not plugged.

Steam Boiler Controls Most low-pressure steam boilers should have two pressure controls; one is the operating control and one is the limit control. The operating control is set for the steam pressure that your building needs, typically 2 pounds of pressure or less. The limit control is set higher than the operating control. The limit control should have a manual reset. This means that the control has a button that has to be pushed to make the boiler operate again. Why would we want this control be a manual reset control? Let us assume that the operating control is set at 2 pounds and the limit control is set at 10 pounds steam pressure. If the boiler pressure raises high enough to trip the limit control (10 PSI), what is that telling you about the operating control? It means that the operating control is not working and may need to be replaced. If the limit control is an automatic reset rather than a manual reset, the boiler could operate on the higher pressure. How long do you think it would take before the higher pressure was discovered? The manual reset forces the owner to look at the boiler in the event of a "No Heat" call. Be sure that you use a siphon or pigtail between the boiler and the pressure control. If the system has a steam to water heat exchanger, the steam pressure may be higher than that, typically 5-6 psig.

An item that will save the future service technician from cursing you is to include a union in the pipe from the siphon to the control. It will allow the technician to replace the control without having to disassemble the entire piping from the boiler to the control. The loop on the pigtail should be perpendicular to the front of the control if using a control with mercury. When steam is applied to the pigtail, it wants to straighten itself. If the loop is sideways, it could affect the pressure setting of the control. The steam siphons should be nonferrous or brass.

Pigtail is installed wrong here. It could affect pressuretrol settings. See arrows. The far right control is piped correct.	This only uses one pigtail for both boilers. It could plug and render both inoperable.

Steam Pressure Gauge Steam boilers require a pressure gauge to display the pressure inside the boiler. To connect the pressure gauge to the steam boiler, you will need to use a siphon or more commonly referred to as a pigtail. The siphon is used to protect the internal diaphragm of the steam pressure gauge from steam. They are also used on steam pressure controls. They operate by allowing the steam to enter the siphon where it will condense. Due to the design of the siphon, the condensate water is trapped inside the siphon and forms a water seal to protect the internal components of the controls and gauges. These should be inspected yearly to make sure they are not plugged. It is sometimes easier to simply replace them than it is to try cleaning the accumulation of mud and dirt from the siphon.

Pressuretrol with Mercury	Pressuretrol without Mercury

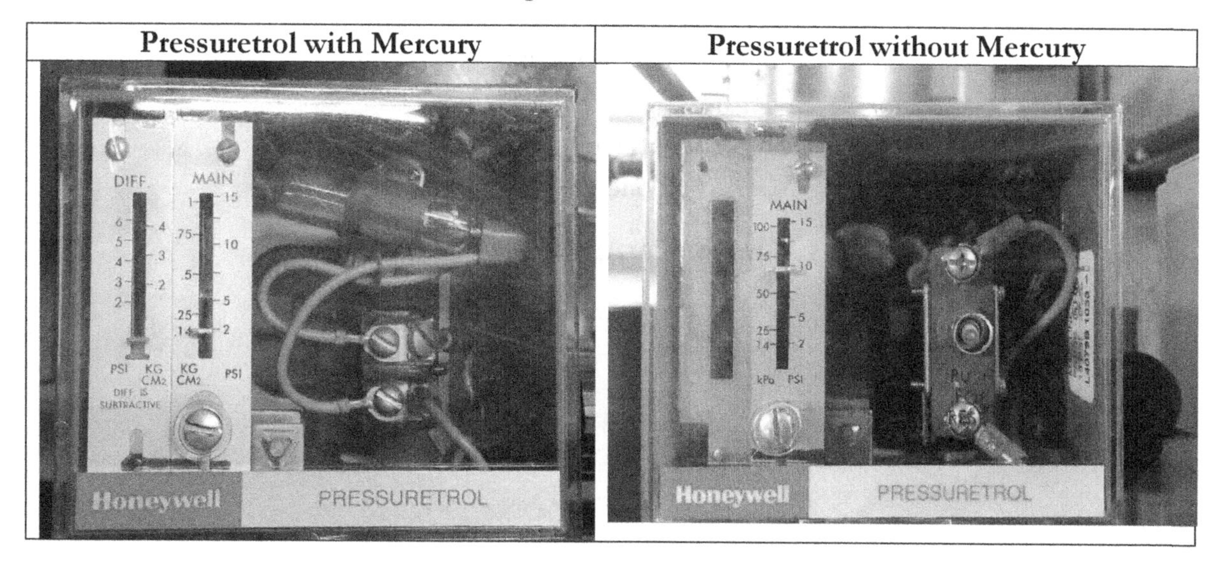

Relief or Safety Valve The relief valve should always be mounted in a vertical position. The rating of the relief valve should be higher than the rated output of the boiler. On steam systems rated at over 500,000 Btuh, the relief valve discharge piping may have to be vented outside. If the relief valve opens, it could fill the room with steam which could make it very dangerous. Steam boiler relief valves use a drip pan elbow on the discharge of the boiler safety valve. The drip pan ell is used to collect condensate that might accumulate in the discharge piping and direct it away from the relief valve. It also limits the piping stress on the relief valve. If venting above the roof, the discharge piping should be 7 feet above the roof and cut at an angle. Remember, the discharge piping from the relief valve cannot be reduced in size. This reduces the relieving capacity of the relief valve, causing a dangerous condition.

Did you know that a firetube boiler has lower maintenance costs than a watertube or cast iron boiler?

Source: ASHRAE

Piping the Steam Boiler

Near Boiler Piping To have a trouble free steam system, the near boiler piping is crucial. The near boiler piping in the picture is incorrect. The steam takeoff to the building is piped between the two risers coming from the boiler. This leads to wet steam and increased operating costs. The pictures below show different ways to pipe a cast iron boiler. On Drawing A, you will notice some extra piping called a "swing joint". This is required on cast iron boilers due to the different rates of expansion of the two metals. The piping would expand at a different rate than the boiler and could result in a damaged boiler. On Drawing B, the swing joint is not installed and this could lead to cracked sections.

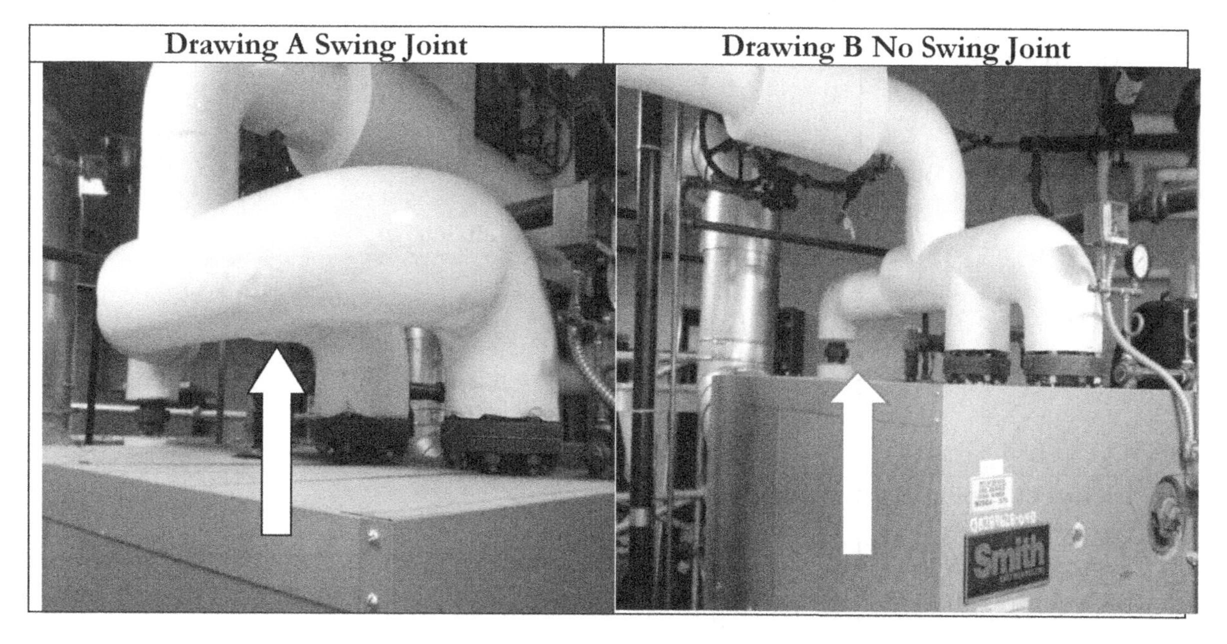

Drawing A Swing Joint | **Drawing B No Swing Joint**

Steam Cast Iron Boiler Header Piping

Near Steam Boiler Piping

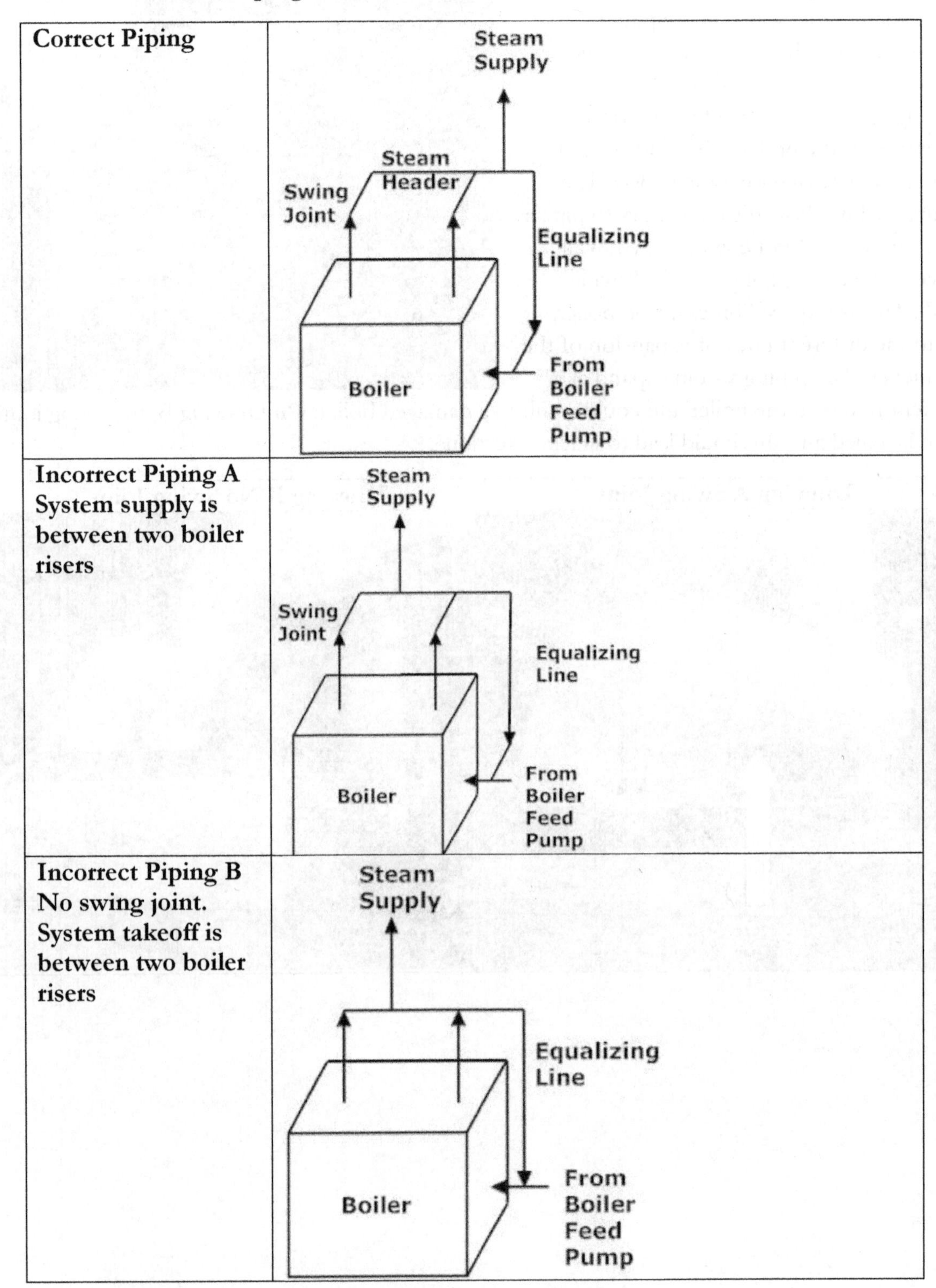

This boiler had massive steam leaks because the boiler piping did not have a swing joint. It pulled the sections apart.	

When piping the steam boiler, the steam header should be a minimum of 24" above the water line. The reducer at the end of the steam main should be installed in the vertical piping. Maintain full pipe diameter until after the turn. This will reduce the chances of wet steam.

Drop Header A drop header is becoming more common as it helps to dry the steam and reduce the chances of carryover. Carryover means that water from the boiler is drawn into the steam piping. This increases the treatment costs and the wet steam could plug controls or valves.

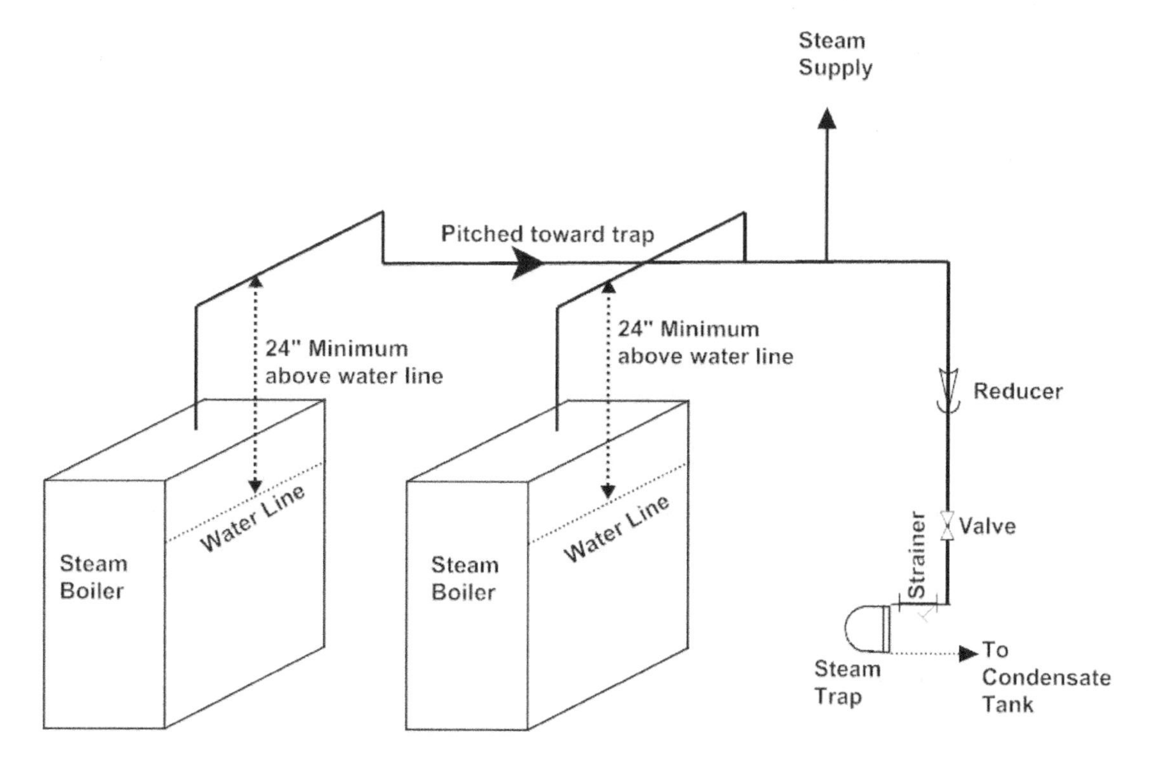

Near Boiler Pipe Size Excessive steam velocity should be avoided in a low-pressure steam system as it will cause water to be entrained in the steam. This is called "Wet Steam." Wet Steam causes a variety of problems.

- Reduced Efficiency – Excessive water in the steam system will cool the steam, causing a drop in system efficiency.
- Increased Maintenance – When the wet steam enters the system, the water causes increased maintenance by plugging the steam valve and trap orifices.
- Increased Water Treatment Costs – When water from the boiler is carried into the system, it carries with it some of the water treatment chemicals. This causes the treatment chemicals to be consumed and will require replacement.

Connecting Boiler to Building Piping When connecting a new steam boiler to the piping, the steam piping should connect to the top of the steam header. If the steam piping is connected to the bottom of the header, the condensate in the piping will drop back into the boiler and could cause flooding. If you cannot pipe the steam into the top of the steam piping, the riser should be trapped so the condensate does not drip back to the boiler.

Correct way to pipe into steam header. This is called a Drop Header.	Incorrect way to pipe into steam header	Alternate way to pipe into the steam header if you cannot pipe into the top of the header

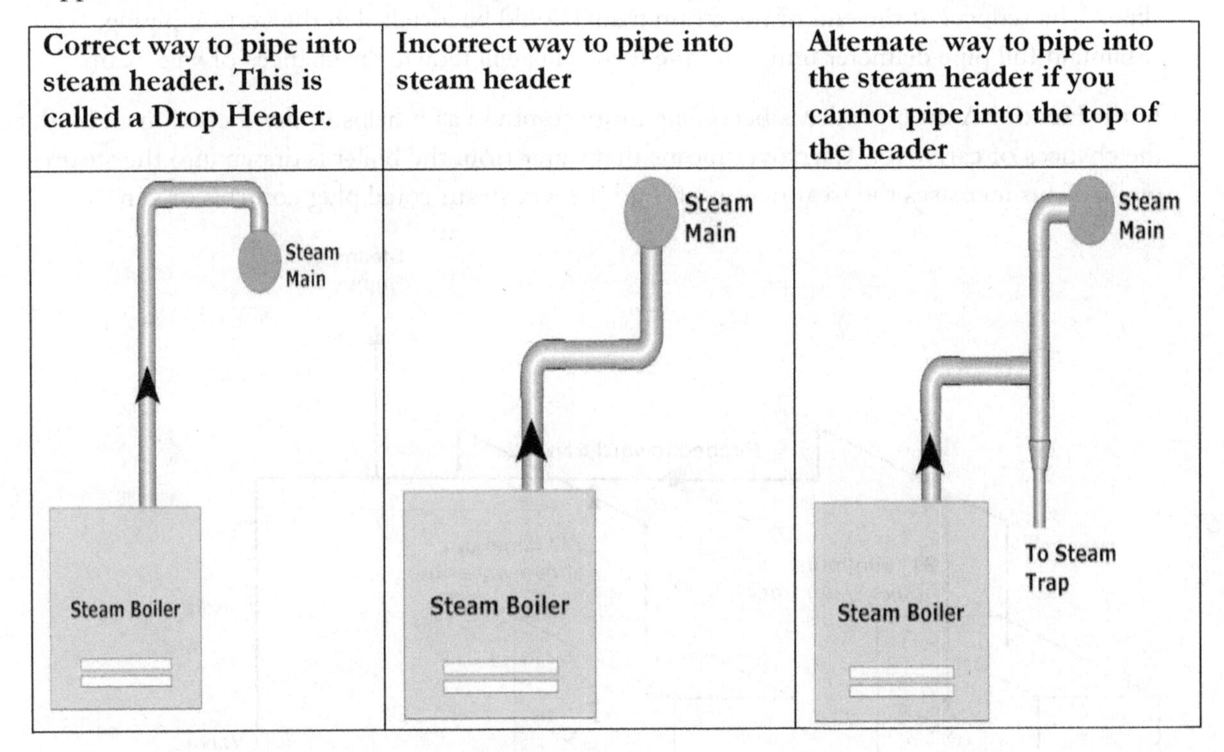

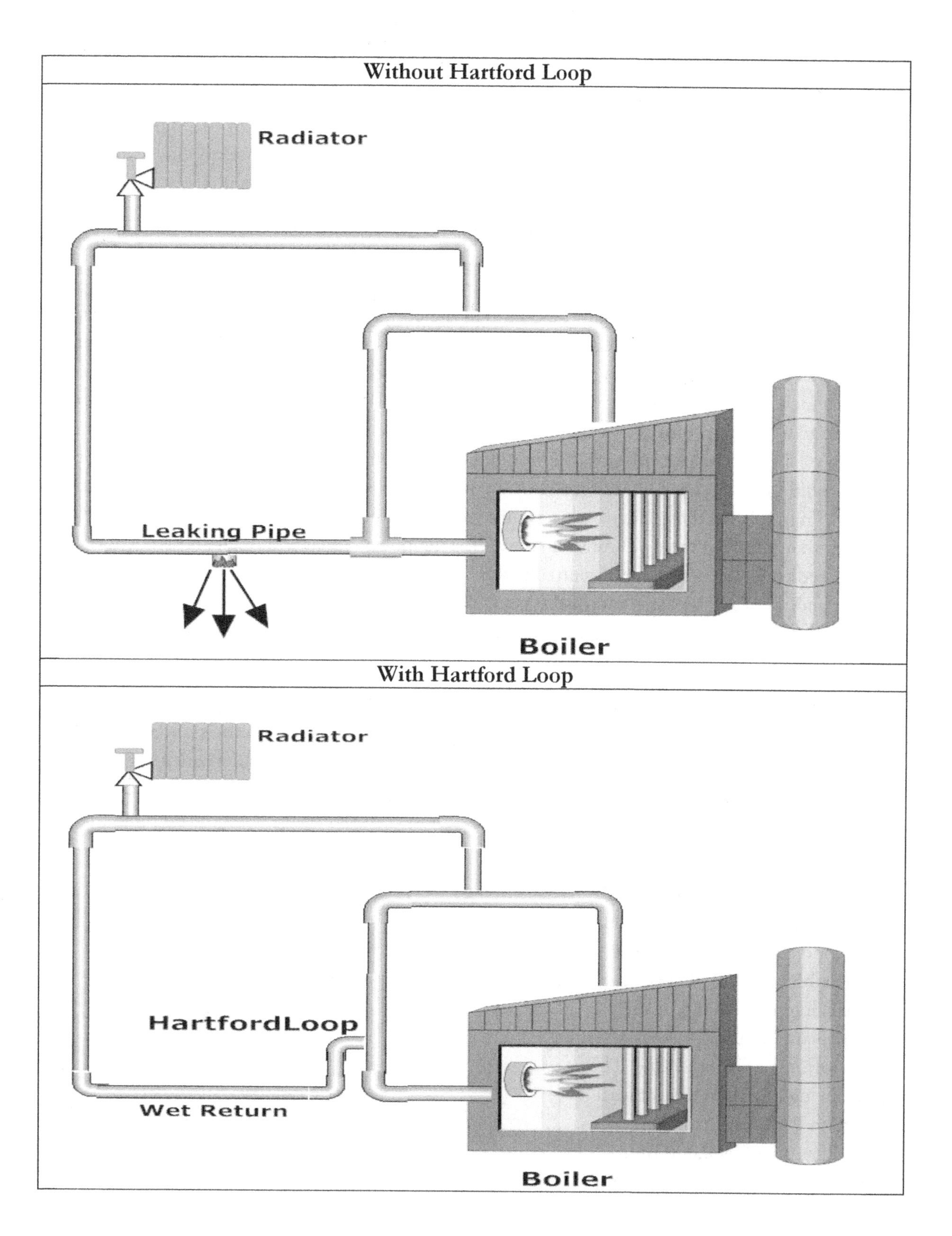

Hartford Loop When coal was used as a fuel source in the old boilers, control was rather difficult. The old coal boilers could not be shut off like the new gas or oil boilers. If there was a leak in the wet return piping, the boiler could lose all its water and dry fire, which is very dangerous. See drawing above, Without Hartford Loop. The Hartford Loop was developed to prevent the water from draining the boiler. With a Hartford loop, a leak on the wet return would not drain the boiler completely of water. It would drop to the elevation of the Hartford Loop fitting. On most commercial boilers with modern safety controls, they are no longer needed. They are still installed on steam boilers without a feed water pump or condensate tank.

Condensate Pipe Size Most steam systems rely on gravity to return the condensate from the system to the condensate tank. Some old steam systems used a vacuum pump to pull the condensate back to the boiler. An advantage to a vacuum return was that the condensate piping could be undersized since the pump would pull the condensate back to the boiler. In addition, the vacuum pump could overcome any of the minor dips in the piping. These vacuum type systems were costly to maintain and most were abandoned in favor of a standard condensate tank, which rely on gravity to return the condensate. The drawback to converting a vacuum system to a gravity return is that the condensate piping may be undersized. This leads to a slow return of the condensate and possible flooding of the condensate tank or boiler. An example of this would be if we size a boiler feed tank for 15 minutes storage. That means that we think it will take 15 minutes for the condensate to return from the system once the boiler is started. If it takes longer than that, the tank will run out of water. The internal float will open and feed fresh, untreated water into the tank. Eventually, the condensate will return. When it does, the tank will flood and the water will spill out of the vent. This will waste the heat as well as the chemicals. If you are unsure as to whether the system used to be a vacuum system, a slightly oversized boiler feed tank will avoid these problems.

Condensate Pipe Note Most of the old steam systems used schedule 80 pipe on the common condensate return piping. Schedule 80 piping is about 50% thicker than the standard schedule 40 pipe. The old designers knew that carbonic acids would form in the condensate lines as a result of the carbon dioxide that mixes with the water in the return piping. If your replacement consists of replacing return piping, remember that you should consider using schedule 80 pipe.

Equalizing Line The equalizing line equalizes the pressure inside the boiler from the steam side to the condensate side. This leads to a stable water level. The equalizing pipe should be the same diameter until it is below the normal boiler water line. The equalizing line should be at least 2 ½" in diameter. If the boiler is rated for over 2,500,000 Btuh, the equalizing line should be at least 4".

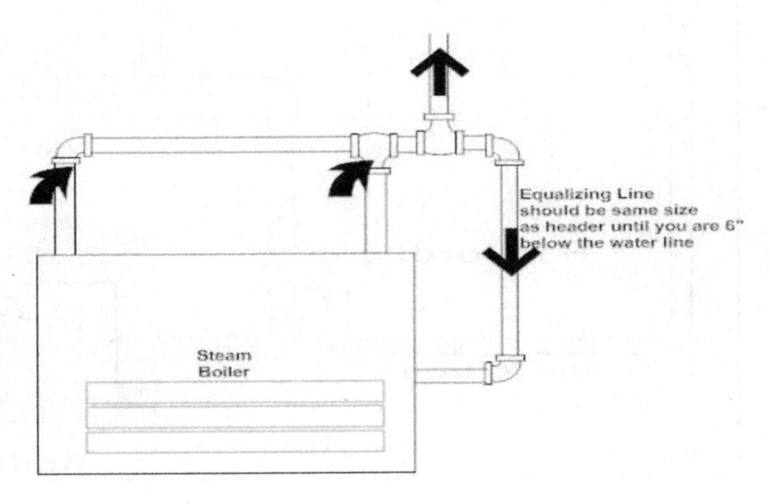

The new boilers subscribe to a different type of philosophy regarding condensate. A boiler feed system is used instead of a condensate tank. The difference between the two is that the boiler feed unit is usually much larger than a condensate tank, typically about five times larger than a condensate tank. The cold makeup water is fed into the boiler feed tank where it will be preheated and chemically treated before being introduced into the boiler. This reduces the chances of thermal shock to the boiler. Another advantage is that the city water, when mixed with the hot condensate, will release some of the oxygen into the boiler feed tank. Since the water level is critical on the new boilers, they incorporate a combination pump control and low water cutoff. When the boiler water level drops to a certain level, a set of contacts inside the pump control located on the boiler will energize the boiler feed pump, filling the boiler.

Simplex Condensate Tank

Duplex Boiler Feed Tank with Spare Pump

In case you were wondering, a duplex boiler feed means that the unit has two pumps.

How to calculate steam velocity

Lbs/Hr x Cubic Volume of Steam divided by 25 x Internal area of pipe = Steam Velocity

$$\text{Steam Velocity} = \frac{Lbs\ Hour \times Cubic\ Volume\ of\ Steam}{25 \times Internal\ Area\ of\ Pipe}$$

Lbs. Steam/Hr = Btuh/960

Lbs. Steam/Hr = Boiler HP x 34.5

Another formula to calculate steam velocity is

$$\text{Steam Velocity} = \frac{2.4\ x\ Lbs\ Hour \times Specific\ Volume\ of\ Steam}{Internal\ Area\ of\ Pipe}$$

Volume of Steam in Cubic Feet per Hour	
PSIG Steam	**Cubic Feet per pound**
0	27
1	25
2	24
3	22
4	21
5	20

Internal Volume of Schedule 40 Pipe	
Pipe Size	**Internal Square Inches**
2"	3.36"
2 ½"	4.78"
3"	7.39"
4"	12.73"
5"	19.99"
6"	28.89"
8"	51.15"
10"	81.55"
12"	114.80"

Steam Velocity One of the leading causes of steam carryover is the near boiler pipe sizing. If the piping is undersized, the steam velocity will be high enough to carry water into the system.

The old timers used to design the near boiler piping for a velocity of 15 fps or feet per second. To keep installation costs down, most new boilers use a higher velocity and smaller pipes. They use the near boiler piping to dry the steam before it enters the system. I performed a study of 50 of the most popular steam boilers and found that the average velocity in the near boiler piping is 45.77 feet per second or fps. Spirax Sarco recommends a maximum velocity of 40 feet per second.

The following table will assure you that the steam velocity will be 40 fps or less, based on 2# of steam pressure. Each boiler manufacturer requires different velocities. Please check their requirements.

Pipe Sizing to Assure Steam Velocities Below 40 FPS		
Pipe Size	**Lbs / Hr**	**Btu/ Hr**
2"	140	134,400
2 ½"	199	191,200
3"	307	295,600
4"	530	509,200
5"	833	799,600
6"	1,204	1,155,600
8"	2,131	2,046,000
10"	3,397	3,262,000
12"	4,783	4,592,000

For example, a boiler rated for 750,000 Btuh output, would require a 5" pipe.

To see the difference between the two designs, here is the capacities if you were designing a steam system with a velocity of 15 feet per second.

Pipe Sizing to Assure Steam Velocities Below 15 FPS		
Pipe Size	**Lbs / Hr**	**Btu/ Hr**
2"	52.5	50,400
2 ½"	74.69	71,700
3"	115.47	110,850
4"	198.91	190,950
5"	312.34	299,850
6"	451.41	433,350
8"	799.22	767,250
10"	1,274.22	1,223,250
12"	1,793.75	1,722,000

For example, a boiler rated for 750,000 Btuh output, would require a 8" pipe

Steam Header Pipe Velocity The above study also showed that the boiler headers had an average velocity of 55 feet per second. The highest velocity was 65 fps and the slowest was 47 fps. The following is a design guide using 50 feet per second.

Pipe Sizing to Assure Steam Velocities Below 50 FPS		
Pipe Size	**Lbs / Hr**	**Btu/ Hr**
2"	175.00	168,000
2 ½"	248.96	239,000
3"	384.90	369,500
4"	663.02	636,500
5"	1,041.15	999,500
6"	1,504.69	1,444,500
8"	2,664.06	2,557,500
10"	4,247.40	4,077,500
12"	5,979.17	5,740,000

How Fast is Fast?

One foot per second = 0.6818 miles per hour. The following is a chart to help you to compare the steam velocities.

Feet per Second to Miles per Hour				
FPS	**MPH**		**FPS**	**MPH**
15	10.2		**50**	34.09
20	13.6		**55**	37.5
25	17		**60**	40.9
30	20.4		**65**	44.31
35	23.86		**70**	47.72
40	27.27		**75**	51.13
45	30.68		**80**	54.54

Trap the Steam Header As steam leaves the boiler and travels throughout the piping, it loses some of its energy. When this happens, some of the steam will condense. To assure dry steam, the main steam supply pipe should trapped at the end of the main. When trapping the header, pipe the header so that it is the full pipe size until you get past the steam supply takeoff to the building and the elbow back to the boiler feed tank. See drawing above You also want to remember the proper static head above the trap.

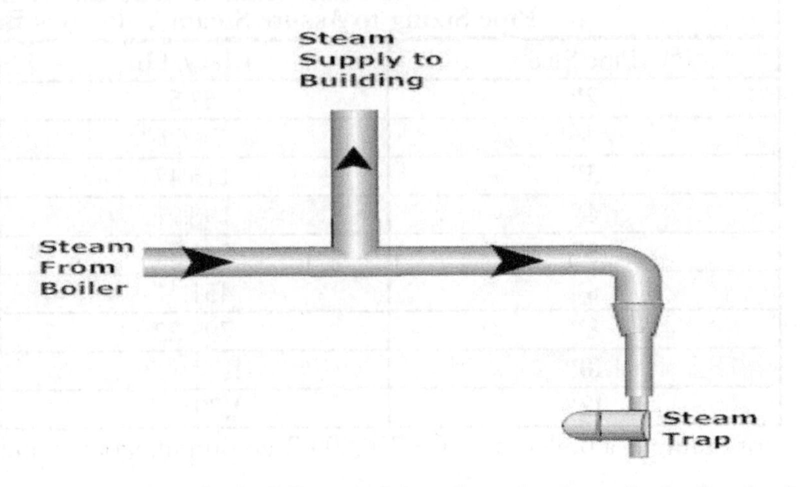

Some near boiler piping rules of thumb

- Steam supply piping to building should always come from the top of the steam piping.
- The steam header should be pitched back toward the equalizing line.

A Couple Thoughts on Steam Traps When trapping the steam header, there are a couple items to remember. The first is that the feed to the trap must be a full size diameter tee that will feed the trap. If you welded a 3/4" tapping into the bottom of a 8" header, most of the water will zoom past the tapping. It would be like driving down the road at 30 miles per hour and reaching out the door to pick up a quarter on the road. In addition, the steam velocity could actually pull the water from the ¾" pipe. See B below.

Another item to remember is that the pipe feeding the steam trap should be at least 15" below the steam header. This will add a static head to the trap to allow it to work better. A 15" static head will provide a 1/2 psi pressure differential across the trap when it drains into a gravity return system that is vented to atmosphere. This is required to drain the trap when the boiler is off and the steam pressure is at 0 psi.

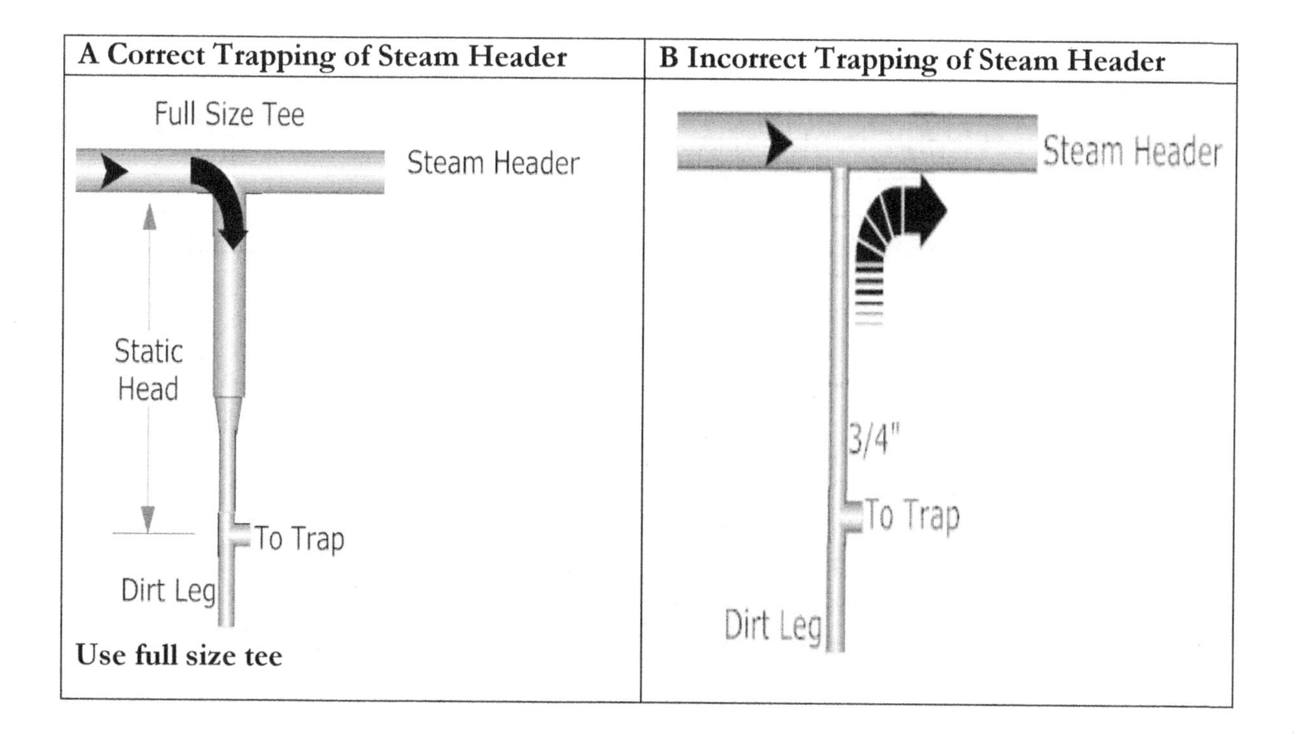

Back Pressure on Steam Trap Capacity. Leaking steam traps will fill the condensate pipe with steam, reducing the pipe's capacity. The steam in the condensate pipe is called "Back Pressure". If the system has steam in the condensate piping, the trap relieving capacity is diminished. This lowers the efficiency of the system. The following chart details the effect of back pressure on the operation of steam traps.

Effect of Back Pressure on Steam Trap Capacity	
Back Pressure	**Loss of Capacity**
25%	6%
50%	20%
75%	38%
Based upon 5# steam pressure	

Air in Steam Lines Air in the steam pipes will reduce the efficiency as well as the temperature of the steam. Properly working air vents could help to eliminate the air. Please see the effects of air in a steam system listed below.

Air in Steam Lines				
Steam Pressure	Pure Steam No Air	5% Air	10% Air	15% Air
2 #	219^{0} F	216^{0} F	213^{0} F	210^{0} F
5 #	227^{0} F	225^{0} F	222^{0} F	219^{0} F
10 #	239^{0} F	237^{0} F	233^{0} F	230^{0} F

Double Trapping When you see a trap on the condensate pipe at the entrance of the condensate or boiler feed tank, be very careful. You will be inheriting a problem job. The problems most likely started when one or two traps started to leak steam. The boiler feed vent would then start spewing steam into the boiler room. Someone then decided to install a "master" trap at the inlet to the condensate or boiler feed tank instead of repairing the traps. That is when the system complaints start. The person then tried raising the steam pressure to overcome the double trapping. The clients fuel bills start to soar and the comfort complaints keep coming in. If you see a trap on the inlet to the boiler feed, you need to inform the client that there are system problems that should be investigated. There are rebuild kits for traps that are relatively inexpensive. This could be an opportunity to get more work from the client.

It's All About Delta P Delta P or pressure drop is the key to how steam systems operate. Pressure goes from high to low always. If there is no pressure drop, the flow stops. Nature hates inequalities and will always try to equalize the system. If you look at the two traps below, the system on the left will allow flow through the trap because of the pressure difference from the inlet to the outlet. The trap on the right will not. The system on the right may have a leaking steam trap in a different area that is pressurizing the condensate side of the system. It will also have steam flowing into the boiler feed or condensate tank.

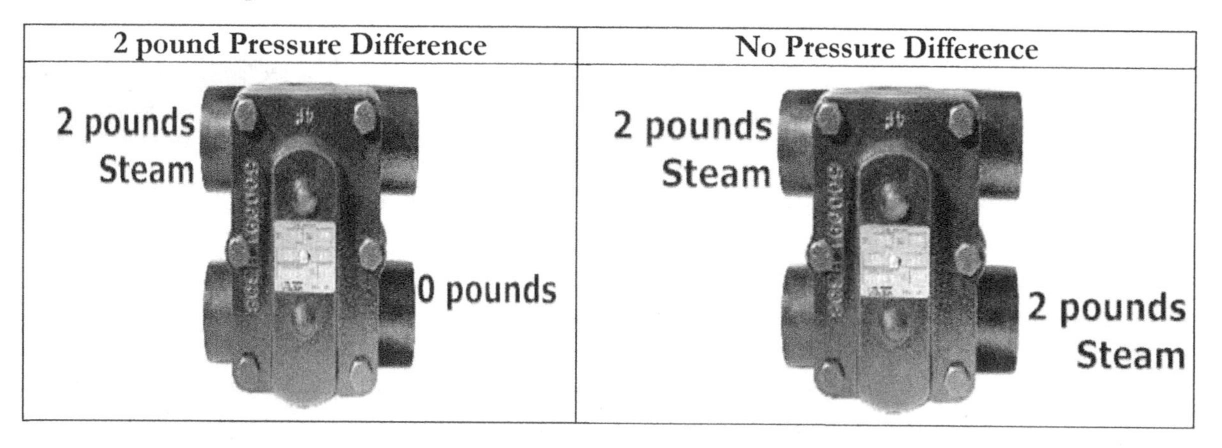

If you consider this ladder type diagram to the right and the trap is leaking through on the bottom radiator, the top two radiators will not have heat as the pressure differential is gone. The bottom radiator will have heat so if you are looking at the top two radiators, you will not find the solution to the problem. Steam system trouble shooting will drive you crazy.

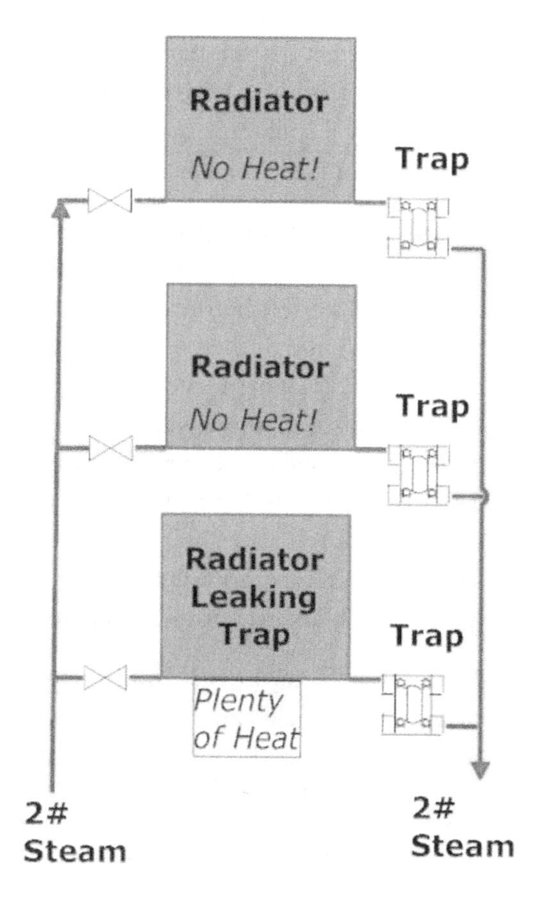

Lag Boiler Flooding A common problem when the heating plant has two or more boilers is that the lag boilers can flood. When this happens, there could be water hammer and in some instances, the boiler will not fire. This occurs because the boilers are connected to a common steam header. The steam will enter the boilers that are not firing and will condense. It will eventually cause the water level inside the idle boiler to rise. If you remember from an earlier chapter, it takes one pound of pressure to cause the water to raise 2.3 feet. Conversely, if we have six feet of water inside the boiler above the pressure control, we will have 2.6 pounds of static pressure on the boiler pressure control. If we are operating the boilers at 2 pounds of pressure, that boiler will never fire until we drain the boiler.

The American Society of Mechanical Engineers or ASME recognized this as a problem and offer the following solution. It is called a "High Level Spill". The High Level Spill is a steam trap that is installed about 1" above the normal water level. If the water level inside the boiler rises to the elevation of the steam trap, the steam trap will open and drain the water from the boiler. The discharge of the trap could be piped to the boiler feed unit so that the water is not wasted. Some water treatment companies do not want the discharge of the high level spill traps piped to the condensate return system as it appears as carry over when they are doing their tests. If the boiler feed tank is far away, a condensate tank could serve as a transfer pump. The common discharge of the high-level spill traps could be piped into a condensate tank that will pump them back to the boiler feed tank.

A High Level Spill Lesson. We were called to a job that was experiencing boiler flooding. The boilers had high level spill traps installed on each boiler. The discharge of each high level spill trap rose three feet before going into the condensate pipe. In addition, there was no check valve on the discharge of the piping after the trap. A check valve is required if you want to "lift" the condensate. The high level spill trap is designed to drain the lag boiler that is not fired. If the boiler is not fired, there will not be enough pressure to lift the condensate, resulting in a flooded boiler. The installer felt that the steam pressure from the other boiler would lift the condensate. The problem with that is that a one ounce difference in steam pressure between the two boilers would result is a water height difference of 1 3/4". In addition, if both boilers were satisfied and shut off, there was no way to lift the condensate and the boilers will flood. If the flooding is too high, the boilers may not start due to the static head on the pressure control.

Bouncing Water Levels Improper chemical treatment, oil, excessive demand, excessive solids, or a vacuum could cause bouncing water levels inside a boiler. Bouncing water levels could cause the boilers to trip the low water cutoff and cause carryover.

If the boilers are chemically treated, check the chemical levels as directed by either the boiler manufacturer or chemical treatment company. The water treatment company can tell you if the solids are elevated in the boiler.

Oil will accumulate in the boilers if they were improperly skimmed or cleaned at startup or if some of the system piping was replaced. When pipe is threaded, the installer uses cutting oil for the pipe threader. This oil can accumulate inside the boiler after several days. Oil is sometimes difficult to remove from a boiler as it adheres to the walls inside the boiler when drained. It may take several attempts to flush the boiler to rid it of the oil. Check with the boiler manufacturer when using a chemical to clean the boiler to verify that it is safe to use on their equipment.

Another cause of bouncing water levels could be zone valves on the steam supply lines. If the zone valves are not installed properly, a vacuum could occur in the piping and could literally suck the water from the boiler when the valve opens. Even if the valves are installed correctly and there is no vacuum, a quick opening valve opening could affect the water levels. If the piping is cool and the valve opens, the steam will rush to the cool piping and cause a surge in the boiler. This could affect the steam velocity and pull the water from the boiler.

We were asked to consult on a steam boiler project where the boiler water level would bounce greatly and trip the low water cutoffs. Some areas of the building would flood and overcome the steam traps, losing heat. After some testing, we found that the water inside the boiler contained high total dissolved solids or TDS. The boiler had carryover, which meant that it was taking water from the boiler and sending it into the steam piping. The carryover was caused by incorrect installation of the near boiler piping. The chemical treatment was an all-inclusive treatment that contained an oxygen scavenger as well as several other chemicals to reduce scale and erosion. The oxygen scavenger removes the oxygen in the boiler to avoid oxygen pitting. When the carry over water from the boiler was transported into the steam and return piping, the air inside the return piping would consume the oxygen scavenger but not the other chemicals. When the maintenance man checked the oxygen scavenger level in the boilers, it would read low. He would then add more chemicals. This led to overfeeding of the chemicals, elevated TDS, bouncing boilers and no heat because the other chemicals in the treatment system were not depleted. To resolve the problem, the near boiler piping had to be replaced. It was an expensive repair.

Checking the Gauge Glass Sometimes, it is difficult to tell by looking at the gauge glass if the boiler is flooded or empty. If you hold a pencil behind the gauge glass, it will give you a clue. If the gauge glass is empty, the pencil will look normal. If the gauge glass is filled with water, the pencil will look broken behind it.

Energy Lost with a Steam Leak

The following is how much energy is lost through a leaking steam trap orifice or hole in a steam pipe.

Steam Pressure 15 psig		
Hole or Orifice Size	**Lbs of Steam lost / Hr**	**Btu's Lost per Hour**
1/8"	18.7	**18,700**
3/16"	42.2	**42,200**
¼"	**75**	**75,000**
5/16"	**117**	**117,000**
3/8"	**168**	**167,000**
7/16"	229	**229,000**
½"	**300**	**300,000**

Steam Pipe Sizing Charts

Steam Pipe Sizes Steam Main

Pipe Size	Btu/Hr		Pipe Size	Btu/Hr
2"	**155,520**		**4"**	**912,000**
2 ½"	**247,680**		**5"**	**1,612,800**
3"	**446,400**		**6"**	**2,707,200**
3 ½"	**643,200**		**8"**	**5,347,200**

Sizing a Dry Return Piping System

Pipe Size	Btu/Hr
1"	**98,800**
1 ¼"	**208,320**
1 ½"	**326,400**
2"	**710,400**
2 ½"	**1,180,800**
3"	**2,160,000**
4"	**4,636,800**

Size Steam Main by Connected Radiation

Radiation Sq Feet	Pipe Size Inches
75-125	1 ¼
125-175	1 ½
175-300	2
300-475	2 ½
475 – 700	3
700-1200	4
1200-1975	5
1975 – 2850	6

Steam Boiler Make Up Water Piping On the makeup piping for the steam boilers, a pressure-reducing valve should be installed in the piping. The internal makeup valve inside the boiler feed tank will sometimes require lower pressure than the city water pressure to the facility. If the water pressure is too high, it will overcome the internal float valve and flood the boiler feed tank.

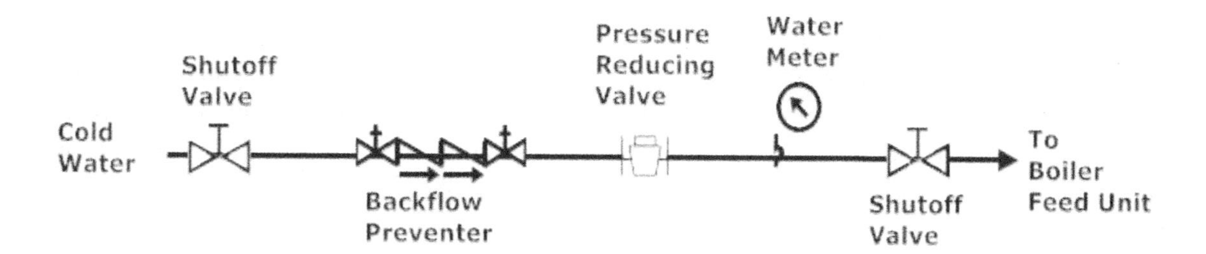

Steam from Boiler Feed Vent If steam is coming from the boiler feed vent pipe, it could indicate a couple problems. The first is that the steam traps could be leaking through. Leaking steam traps could allow steam to escape and makeup water would have to be introduced. Leaking steam traps could also cause some areas to have comfort complaints. The second cause could be excessive steam pressure. Float and thermostatic steam traps are designed to open once the temperature drops 20 degrees F. If the boiler is set at 7# steam pressure, the temperature of the steam is 232 degrees F. That means that the condensate will be discharged into the pipe at 212 degrees F. As you are aware, that is the boiling temperature of water at atmospheric pressure so the condensate could flash to steam. When water turns to steam, it expands 1,600 times its volume. The flow of condensate back to the boiler slows and the pipe is filled with steam. This boiler feed unit runs out of water and feeds fresh, untreated water into the system. It also increases the temperature inside the condensate tank. Elevated condensate temperatures could cause damage to the boiler feed pump. If the water is too warm, the condensate could flash to steam inside the pump vortex and destroy it. This is called cavitation and it sounds like the pump has marbles inside it. The steam will destroy the pump's impeller and reduce pump capacity. In many condensate pumps, the impeller is made of a softer metal like bronze. It could also attack and destroy the pump's mechanical seal. Many pumps are designed for operation under 200 degree F.

Condensate Tank or Boiler Feed Many of the older steam systems had the system makeup water piped into the boiler. The industry has rethought that as the cold makeup water could cause thermal shock to the boiler. If the system you are working on has the makeup water feeding the boiler, it most likely used a condensate tank. A condensate tank has a smaller tank than a boiler feed unit, typically around 20 gallons. It has an internal float that will operate the pump. The pump will operate anytime that the float is high enough inside the tank, regardless of the water level in the boiler.

Condensate Tank The condensate from the building will accumulate inside the tank. The condensate tank is smaller than a boiler feed tank, usually around 15-25 gallons. When the condensate water level raises high enough, the internal float switch energizes the pump and feeds water into the boiler, regardless of whether the boiler needs water or not. The pump will operate until the water level inside the tank drops low enough to open the float switch contacts. Any makeup water required by the steam system is fed through a level control located on the boiler. The makeup water for the system is fed directly into the boiler via a float-operated control. This could be problematic as cold water entering a hot boiler can cause many problems.

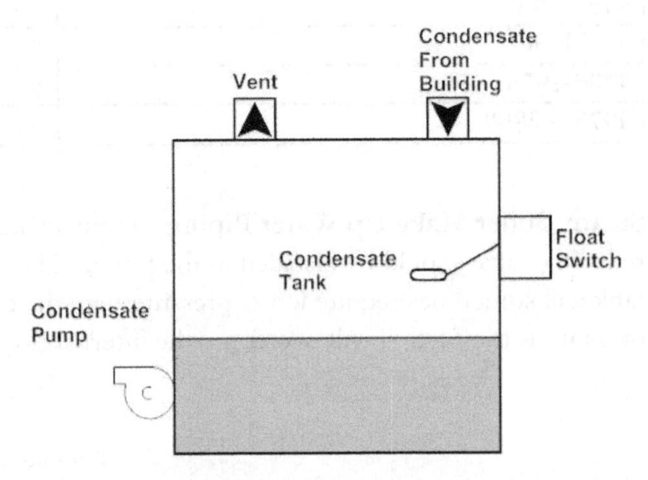

Boiler Feed Unit Condensate from the building will accumulate inside the boiler feed tank. The boiler feed tank is much larger than a condensate tank. When the condensate water level drops below the makeup water float, the valve attached to the makeup water float will open, feeding fresh water into the tank. As the tank water level rises, the makeup water float valve will close and stop feeding water into the tank. Operation of the pump is controlled by a combination pump control / low water cutoff, such as a McDonnell Miller #150, that is located on the boiler. As the boiler water level drops, the pump will be energized. The pump will operate until the boiler water level control is satisfied. Makeup water for the system is introduced into the boiler feed tank where it will be pre-heated.

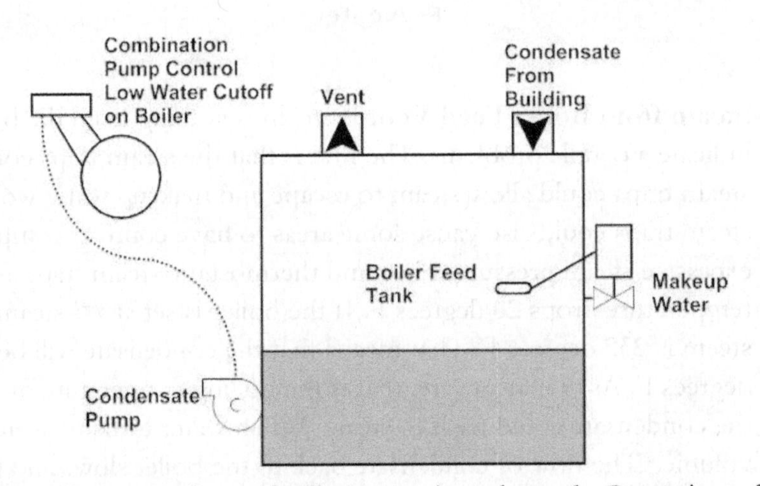

Boiler Feed Sizing In all steam boilers, the steaming rate is the same, regardless of style or construction. All steam boilers will convert the boiler water to steam at approximately one half gpm per 240,000 gross output Btu/Hr (D.O.E Heating capacity). A rule of thumb in the industry states that we need One gpm per 240,000 makeup to the boiler. This is so that the boiler does not run out of water.

Sizing a boiler feed tank is not an exact science and is based on the time it takes the system from a cold start until when the condensate returns to the tank. The most accurate way to do this is to measure the time it takes from when steam is generated until the condensate returns from the system. In most instances, designers size the boiler feed storage tank for anywhere from 10 to 30 minutes. If you have a long building, you may want to increase the tank size as the return could take longer. Some manufacturers suggest a 50% safety factor when sizing a tank. In reality, there is usually only a small amount of cost difference between one tank size and the next. You are better off with a slightly oversized tank than an undersized one.

Things to know

- BHP = Boiler Horse Power
- One Boiler HP = 34.5 lbs of steam/ hr from and at 212 degrees F.
- One gallon of water weighs approximately 8.337 pounds.
- 10 – 30 minutes storage time = rule of thumb for water storage for a boiler feed tank. This is the length of time it takes for condensate to return from the building.
- Multiply tank size by 1.5 for a safety factor.
- GPM = EDR x 0.000496
- GPM = BTU ÷ 480,000
- Pounds of Condensate/Hr = EDR ÷ 4
- To calculate boiler evaporation rate in gallons per minute GPM = BHP x .069 Example 100 BHP x 0.069 = 6.9 GPM

Calculate Storage Tank Sizing

Boiler Horsepower (BHP) x 34.5 / 8.337 lbs /60 minutes x Storage Time x (1.5) safety factor. This will give you a 50% larger tank for a safety factor.

For example 10 minute storage:

100 BHP x 34.5 / 8.337/ 60 x 10 = 68.9 gallons

Multiply 68.9 x 1.5 (Safety Factor) = 103.35 Gallon Tank.

If you choose a tank with a capacity greater than 103 gallons, that would be large enough for this project. This is based upon 10 minutes storage.

For example 20 minute storage:

100 BHP x 34.5 / 8.337/ 60 x 20 = 137.8 gallons

Multiply 137.8 x 1.5 (Safety Factor) = 206.7 Gallon Tank.

If you choose a tank with a capacity greater than 206 gallons, that would be large enough for this project. This is based upon 20 minutes storage.

A larger storage tank may be required if the condensate is slow to return. A safety factor is required because of possible system deficiencies and the fact that some of the tank is below the pump inlet.

A couple other rules of thumb for tank sizing are:

- Boiler evaporation rate in GPM x 20 = Tank Size. (This is based on 10 minute storage)
- Boiler evaporation rate in GPM x 40 = Tank Size. (This is based on 20 minute storage)
- Boiler evaporation rate in GPM x 60 = Tank Size. (This is based on 30 minute storage)
- One gallon of storage for each boiler HP for a small building. Two gallons of storage for each boiler hp for a larger building.

Boiler Feed Water Pump Sizing To calculate the pump capacity, you will need to know the evaporation rate of the boiler. You should then add a 50% to 100% safety margin.

To calculate the boiler evaporation rate, please use the following formula:

Evaporation rate = Boiler HP x 34.5 / 8.337(lbs) / 60 (minutes)

For Example:

BHP x 34.5 / 8.337 / 60 minutes

100 BHP x 34.5 / 8.337 / 60 = 6.89 Gpm (Evaporation Rate) x 1.5 (50% Safety Factor) = 10.33 Gpm Pump

The following are some rules of thumb for boiler feed pump sizing

- 1/10 gpm per Boiler HP or BHP
  E.g. 100 hp = 10 gpm

- times boiler maximum evaporation rate or 0.14 GPM per boiler HP for intermittent operation.

- 1.5 times boiler maximum evaporation rate or 0.104 GPM per boiler HP for continuous operation.

Pump Discharge Pressure

The pump discharge pressure should be 3% higher than the relief valve setting plus pressure drop. Always install a valve on the discharge of the pump to limit the feed water pressure.

For example:

15 psig relief valve setting and 2 pound pressure drop

15 x 1.03 + 2 = 17.45 pounds. Your pump will have to have a discharge pressure of at least 17.45 pounds.

Therefore, our system will consist of the following

Tank Size - 103 gallon (10 Minutes)

Or

206 gallon (20 Minutes)

Pump GPM - 10 GPM

Pump Discharge 17.45 pounds

Another Sizing Option

100 BHP Gross Output = 3,347,200 Btuh

3,347,200 / 970 Btu = 3,451 Lbs of steam evaporated per hour.

The evaporation rate is 3,451 / 8.33 = 414 gallons per hour (gph) or 6.90 gallons per minute GPM. Multiply 6.90 x 1.5 (50%) safety factor) for the pump GPM size.

Tank size 414 / 3(20 minutes before condensate begins to return) = 138 gallons. Multiply x 1.5 for safety factor. This equals tank size of 207 gallons.

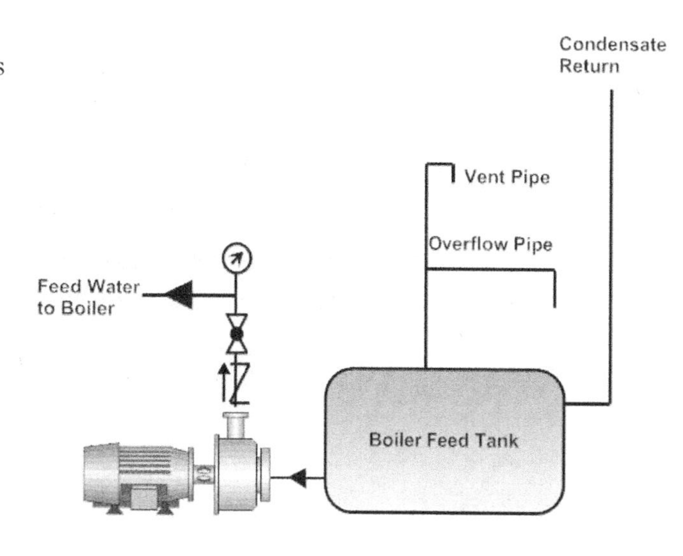

Condensate Tank & Pump Sizing

Evaporation rate x 3 = Pump GPM required

Pump GPM x 1 = Tank sizing

Please note that most new boilers require a boiler feed instead of a condensate tank.

If you want a chart that calculates the sizing, please see below.

How Many Pumps? When I design a steam system with a new boiler feed unit, I prefer using one boiler feed pump per boiler. I prefer this to using makeup water valves on each boiler. The drawback is that if one pump fails, the boiler also is inoperable.

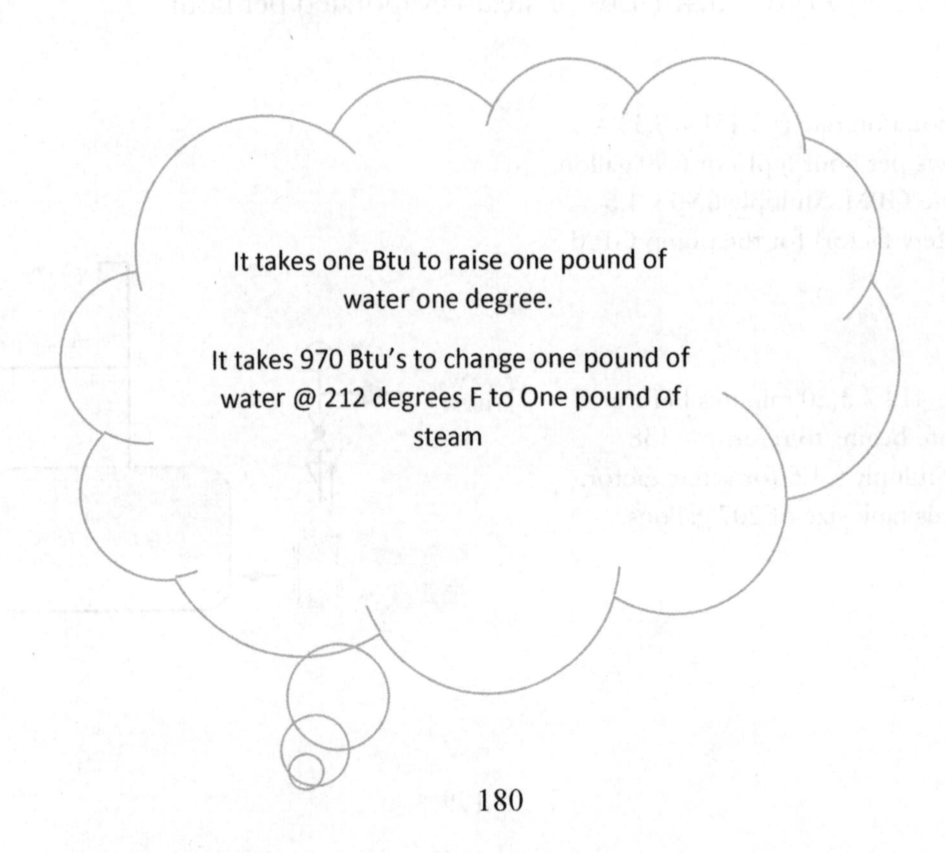

Boiler Feed Tank Sizing				
Boiler HP	10 Minutes	20 Minutes	30 Minutes	Pump GPM
20	21	41	62	2.07
30	31	62	93	3.10
40	41	83	124	4.14
50	52	103	155	5.17
60	62	124	186	6.21
70	72	145	217	7.24
80	83	166	248	8.28
90	93	186	279	9.31
100	103	207	310	10.35
110	114	228	341	11.38
120	124	248	372	12.41
130	134	269	403	13.45
140	145	290	435	14.48
150	155	310	466	15.52
160	166	331	497	16.55
170	176	352	528	17.59
180	186	372	559	18.62
190	197	393	590	19.66
200	207	414	621	20.69
250	259	517	776	25.86
300	310	621	931	31.04
350	362	724	1,086	36.21
400	414	828	1,241	41.38
Tank & pump sizing based upon 50% safety factor				

Trouble Shooting a Condensate Return System

1 Steam is coming from the vent. This could indicate that either the steam pressure is excessive or there are leaking steam traps.

2 Condensate pump sounds noisy. This could be from the water temperature is too high and it is flashing to steam inside the pump volute or the impeller is defective

3 Condensate tank overflows. It could mean that there are issues with the returning condensate or the tank could be too small.

Steam Trap Primer

The following was provided by Engineeringtoolbox.com

Inverted Bucket Steam Trap The inverted bucket is the most reliable steam trap operating principle known. The heart of its simple design is a unique leverage system that multiplies the force provided by the bucket to open the valve against pressure. Since the bucket is open at the bottom, it resists damage from water hammers, and wearing points are heavily reinforced for long life.

- intermittent operation - condensate drainage is continuous, discharge is intermittent
- small dribble at no load, intermittent at light and normal load, continuous at full load
- excellent energy conservation
- excellent resistance to wear
- excellent corrosion resistance
- excellent resistance to hydraulic shocks
- vents air and CO_2 at steam temperature
- poor ability to vent air at very low pressure
- fair ability to handle start up air loads
- excellent operation against back pressure
- good resistance to damage from freezing
- excellent ability to purge system
- excellent performance on very light loads
- immediate responsiveness to slugs of condensate
- excellent ability to handle dirt
- large comparative physical size
- fair ability to handle flash steam
- open at mechanical failure

Thermostatic Steam Traps There are two basic designs for the thermostatic steam trap, a bimetallic and a balanced pressure design. Both designs use the difference in temperature between live steam and condensate or air to control the release of condensate and air from the steam line. In a thermostatic bimetallic trap it is common that an oil filled element expands when heated to close a valve against a seat. It may be possible to adjust the discharge temperature of the trap - often between 60°C and 100°C. This makes the thermostatic trap suited to get rid of large quantities of air and cold condensate at the start-up condition. On the other hand the thermostatic trap will have problems to adapt to the variations common in modulating heat exchangers.

- intermittent operation
- fair energy conservation
- fair resistance to wear
- good corrosion resistance
- poor resistance to hydraulic shocks (good for bimetal traps)
- do not vent air and CO_2 at steam temperature
- good ability to vent air at very low pressure
- excellent ability to handle start up air loads
- excellent operation against back pressure
- good resistance to damage from freezing
- good ability to purge system
- excellent performance on very light loads
- delayed responsiveness to slugs of condensate
- fair ability to handle dirt
- small comparative physical size
- poor ability to handle flash steam
- open or closed at mechanical failure depending of the construction

Float Steam Traps

In the float steam trap a valve is connected to a float in such a way that a valve opens when the float rises. The float steam trap adapts very well to varying conditions as is the best choice for modulating heat exchangers, but the float steam trap is relatively expensive and not very robust against water hammers.

- continuous operation but may cycle at high pressures
- no action at no load, continuous at full load
- good energy conservation
- good resistance to wear
- good corrosion resistance
- poor resistance to hydraulic shocks
- do not vent air and CO_2 at steam temperature
- excellent ability to vent air at very low pressure
- excellent ability to handle start up air loads
- excellent operation against back pressure
- poor resistance to damage from freezing
- fair ability to purge system
- excellent performance on very light loads
- immediate responsiveness to slugs of condensate
- poor ability to handle dirt
- large comparative physical size
- poor ability to handle flash steam
- closed at mechanical failure

Thermodynamic Disc Steam Traps

The thermodynamic trap is an robust steam trap with simple operation. The trap operates by means of the dynamic effect of flash steam as it passes through the trap.

- intermittent operation
- poor energy conservation
- poor resistance to wear
- excellent corrosion resistance
- excellent resistance to hydraulic shocks
- do not vent air and CO_2 at steam temperature
- not recommended at low pressure operations
- poor ability to handle start up air loads
- poor operation against back pressure
- good resistance to damage from freezing
- excellent ability to purge system
- poor performance on very light loads
- delayed responsiveness to slugs of condensate
- poor ability to handle dirt
- small comparative physical size
- poor ability to handle flash steam

Steam trap inserts can be replaced on steam traps to renew them	

Pipe Sag

The following shows a sag in the condensate line from the trap. If that occurs, the air cannot be emptied from the system creating a comfort complaint.

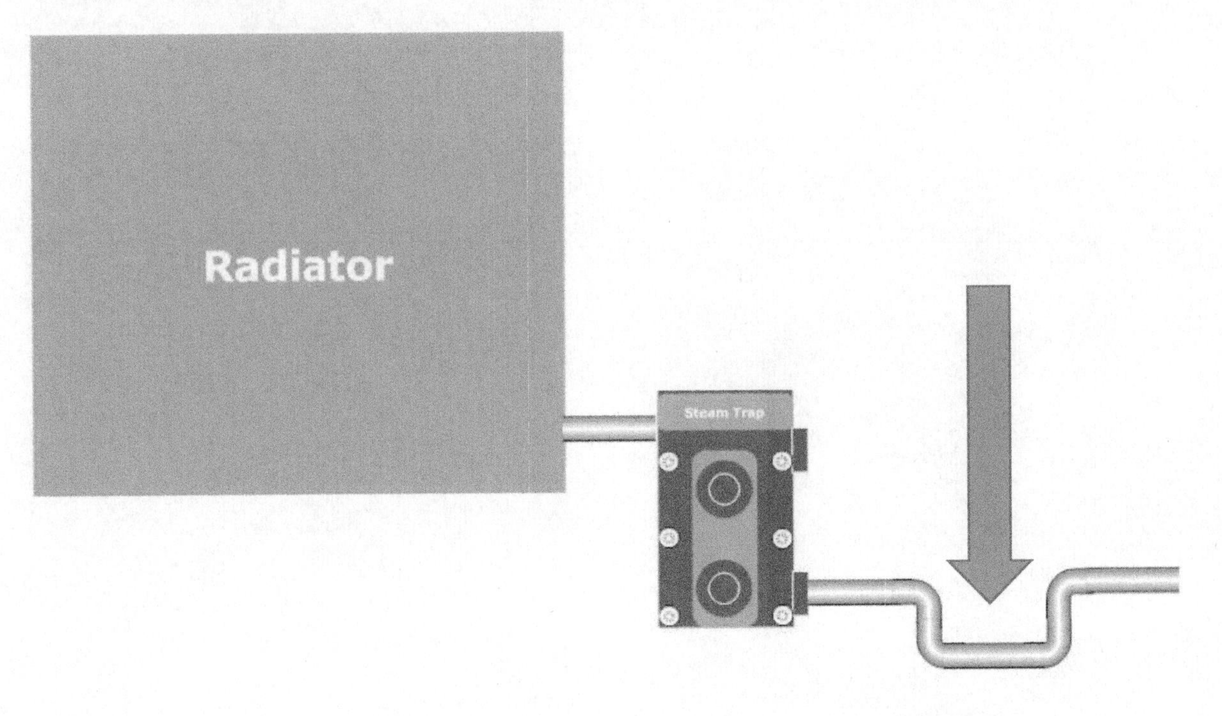

Gauge Glasses

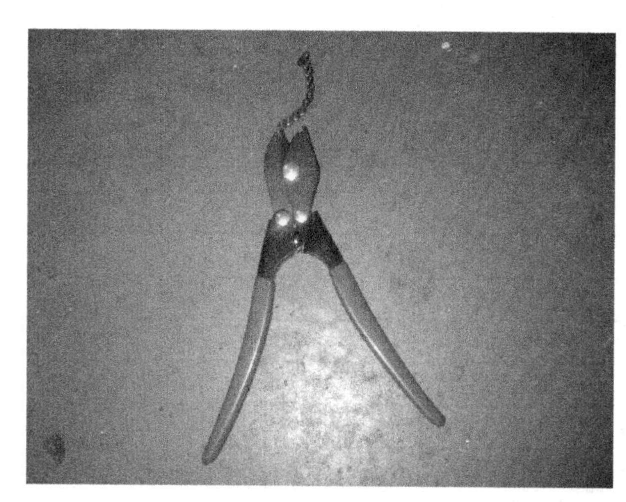

In some instances, you may have to replace the gauge glass on the boiler or compression tank. They gauge glasses usually come in different lengths such as 12", 18" and 24". There are special tools that are available to make the job easier, called a Gauge Glass Cutter. They can be purchased at the same place that you purchase boilers. If you do not have one, a file can be used to trim the gauge glass. It is time consuming and very difficult to do. To use a file, you have to keep running the file over the line that marks the desired length of the gauge glass. After the glass is etched, the gauge glass can be tapped to break it. In most instances, the gauge glass does not break evenly. When installing the gauge glass onto the boiler, slide the brass nut onto the gauge glass. Then slide the brass washer followed by the rubber gasket. The brass washer and rubber gasket should fit inside the brass nut. When tightening the nuts onto the fittings, they should only be tightened until they do not leak. Do not over-tighten as they could break.

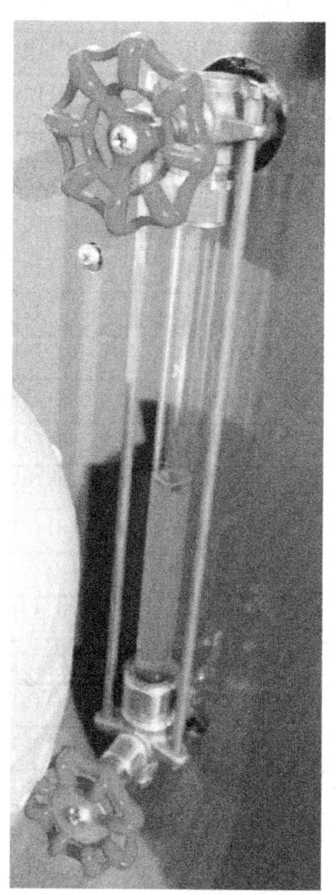

Steam Velocity @ 2 PSIG								
Steam Pressure	**2 PSIG**	Pipe Size						
		2	2 1/2	4	6	8	10	12
Boiler Input	Boiler Output	**Feet Per Second Shaded cells are greater than 40 feet per second**						
200,000	160,000	47	33	21	5	3	2	1
300,000	240,000	71	49	32	8	5	3	2
400,000	320,000	94	66	43	11	6	4	3
500,000	400,000	118	82	53	14	8	5	3
600,000	480,000	141	99	64	16	9	6	4
700,000	560,000	165	115	75	19	11	7	5
800,000	640,000	188	132	86	22	12	8	6
900,000	720,000	212	148	96	25	14	9	6
1,000,000	800,000	236	165	107	27	15	10	7
1,100,000	880,000	259	181	118	30	17	11	8
1,200,000	960,000	283	198	128	33	19	12	8
1,300,000	1,040,000	306	214	139	36	20	13	9
1,400,000	1,120,000	330	231	150	38	22	14	10
1,500,000	1,200,000	353	247	160	41	23	15	10
1,600,000	1,280,000	377	264	171	44	25	16	11
1,700,000	1,360,000	401	280	182	47	26	16	12
1,800,000	1,440,000	424	297	193	49	28	17	12
1,900,000	1,520,000	448	313	203	52	29	18	13
2,000,000	1,600,000	471	330	214	55	31	19	14
2,500,000	2,000,000	589	412	267	69	39	24	17
3,000,000	2,400,000	707	495	321	82	46	29	21
3,500,000	2,800,000	825	577	374	96	54	34	24
4,000,000	3,200,000	942	660	428	110	62	39	28
4,500,000	3,600,000	1,060	742	481	123	70	44	31
5,000,000	4,000,000	1,178	825	535	137	77	49	34
5,500,000	4,400,000	1,296	907	588	151	85	53	38
6,000,000	4,800,000	1,414	989	642	164	93	58	41
6,500,000	5,200,000	1,531	1,072	695	178	101	63	45
7,000,000	5,600,000	1,649	1,154	749	192	108	68	48
7,500,000	6,000,000	1,767	1,237	802	206	116	73	52
8,000,000	6,400,000	1,885	1,319	856	219	124	78	55
8,500,000	6,800,000	2,003	1,402	909	233	132	82	59
9,000,000	7,200,000	2,120	1,484	963	247	139	87	62
9,500,000	7,600,000	2,238	1,567	1,016	260	147	92	66
10,000,000	8,000,000	2,356	1,649	1,070	274	155	97	69

Steam Velocity @ 4 PSIG								
Steam Pressure	**4 PSIG**	Pipe Size						
		2	2 1/2	4	6	8	10	12
Boiler Input	Boiler Output	**Feet Per Second Shaded cells are greater than 40 feet per second**						
200,000	160,000	43	30	20	5	3	2	1
300,000	240,000	65	45	29	8	4	3	2
400,000	320,000	86	60	39	10	6	4	3
500,000	400,000	108	76	49	13	7	4	3
600,000	480,000	130	91	59	15	9	5	4
700,000	560,000	151	106	69	18	10	6	4
800,000	640,000	173	121	78	20	11	7	5
900,000	720,000	194	136	88	23	13	8	6
1,000,000	800,000	216	151	98	25	14	9	6
1,100,000	880,000	238	166	108	28	16	10	7
1,200,000	960,000	259	181	118	30	17	11	8
1,300,000	1,040,000	281	197	127	33	18	12	8
1,400,000	1,120,000	302	212	137	35	20	12	9
1,500,000	1,200,000	324	227	147	38	21	13	9
1,600,000	1,280,000	346	242	157	40	23	14	10
1,700,000	1,360,000	367	257	167	43	24	15	11
1,800,000	1,440,000	389	272	177	45	26	16	11
1,900,000	1,520,000	410	287	186	48	27	17	12
2,000,000	1,600,000	432	302	196	50	28	18	13
2,500,000	2,000,000	540	378	245	63	35	22	16
3,000,000	2,400,000	648	454	294	75	43	27	19
3,500,000	2,800,000	756	529	343	88	50	31	22
4,000,000	3,200,000	864	605	392	101	57	36	25
4,500,000	3,600,000	972	680	441	113	64	40	28
5,000,000	4,000,000	1,080	756	490	126	71	44	32
5,500,000	4,400,000	1,188	831	539	138	78	49	35
6,000,000	4,800,000	1,296	907	588	151	85	53	38
6,500,000	5,200,000	1,404	983	637	163	92	58	41
7,000,000	5,600,000	1,512	1,058	686	176	99	62	44
7,500,000	6,000,000	1,620	1,134	735	188	106	67	47
8,000,000	6,400,000	1,728	1,209	784	201	113	71	51
8,500,000	6,800,000	1,836	1,285	833	214	121	76	54
9,000,000	7,200,000	1,944	1,361	883	226	128	80	57
9,500,000	7,600,000	2,052	1,436	932	239	135	85	60
10,000,000	8,000,000	2,160	1,512	981	251	142	89	63

Steam Velocity @ 6 PSIG								
Steam Pressure	6 PSIG	Pipe Size						
		2	2 1/2	4	6	8	10	12
Boiler Input	Boiler Output	Feet Per Second Shaded cells are greater than 40 feet per second						
200,000	160,000	57	40	26	7	4	2	2
300,000	240,000	85	60	39	10	6	4	2
400,000	320,000	113	79	52	13	7	5	3
500,000	400,000	142	99	64	16	9	6	4
600,000	480,000	170	119	77	20	11	7	5
700,000	560,000	199	139	90	23	13	8	6
800,000	640,000	227	159	103	26	15	9	7
900,000	720,000	255	179	116	30	17	11	7
1,000,000	800,000	284	199	129	33	19	12	8
1,100,000	880,000	312	218	142	36	20	13	9
1,200,000	960,000	340	238	155	40	22	14	10
1,300,000	1,040,000	369	258	167	43	24	15	11
1,400,000	1,120,000	397	278	180	46	26	16	12
1,500,000	1,200,000	425	298	193	49	28	18	12
1,600,000	1,280,000	454	318	206	53	30	19	13
1,700,000	1,360,000	482	337	219	56	32	20	14
1,800,000	1,440,000	510	357	232	59	34	21	15
1,900,000	1,520,000	539	377	245	63	35	22	16
2,000,000	1,600,000	567	397	258	66	37	23	17
2,500,000	2,000,000	709	496	322	82	47	29	21
3,000,000	2,400,000	851	596	386	99	56	35	25
3,500,000	2,800,000	993	695	451	115	65	41	29
4,000,000	3,200,000	1,134	794	515	132	75	47	33
4,500,000	3,600,000	1,276	893	579	148	84	53	37
5,000,000	4,000,000	1,418	993	644	165	93	58	42
5,500,000	4,400,000	1,560	1,092	708	181	102	64	46
6,000,000	4,800,000	1,702	1,191	773	198	112	70	50
6,500,000	5,200,000	1,843	1,290	837	214	121	76	54
7,000,000	5,600,000	1,985	1,390	901	231	130	82	58
7,500,000	6,000,000	2,127	1,489	966	247	140	88	62
8,000,000	6,400,000	2,269	1,588	1,030	264	149	93	66
8,500,000	6,800,000	2,411	1,687	1,095	280	158	99	71
9,000,000	7,200,000	2,552	1,787	1,159	297	168	105	75
9,500,000	7,600,000	2,694	1,886	1,223	313	177	111	79
10,000,000	8,000,000	2,836	1,985	1,288	330	186	117	83

Steam Velocity @ 8 PSIG								
Steam Pressure	8 PSIG	Pipe Size						
		2	2 1/2	4	6	8	10	12
Boiler Input	Boiler Output	**Feet Per Second Shaded cells are greater than 40 feet per second**						
200,000	160,000	35	25	16	4	2	1	1
300,000	240,000	53	37	24	6	3	2	2
400,000	320,000	71	49	32	8	5	3	2
500,000	400,000	88	62	40	10	6	4	3
600,000	480,000	106	74	48	12	7	4	3
700,000	560,000	124	87	56	14	8	5	4
800,000	640,000	141	99	64	16	9	6	4
900,000	720,000	159	111	72	19	10	7	5
1,000,000	800,000	177	124	80	21	12	7	5
1,100,000	880,000	194	136	88	23	13	8	6
1,200,000	960,000	212	148	96	25	14	9	6
1,300,000	1,040,000	230	161	104	27	15	9	7
1,400,000	1,120,000	247	173	112	29	16	10	7
1,500,000	1,200,000	265	186	120	31	17	11	8
1,600,000	1,280,000	283	198	128	33	19	12	8
1,700,000	1,360,000	300	210	136	35	20	12	9
1,800,000	1,440,000	318	223	144	37	21	13	9
1,900,000	1,520,000	336	235	152	39	22	14	10
2,000,000	1,600,000	353	247	160	41	23	15	10
2,500,000	2,000,000	442	309	201	51	29	18	13
3,000,000	2,400,000	530	371	241	62	35	22	16
3,500,000	2,800,000	618	433	281	72	41	25	18
4,000,000	3,200,000	707	495	321	82	46	29	21
4,500,000	3,600,000	795	557	361	93	52	33	23
5,000,000	4,000,000	883	618	401	103	58	36	26
5,500,000	4,400,000	972	680	441	113	64	40	28
6,000,000	4,800,000	1,060	742	481	123	70	44	31
6,500,000	5,200,000	1,149	804	521	134	75	47	34
7,000,000	5,600,000	1,237	866	562	144	81	51	36
7,500,000	6,000,000	1,325	928	602	154	87	55	39
8,000,000	6,400,000	1,414	989	642	164	93	58	41
8,500,000	6,800,000	1,502	1,051	682	175	99	62	44
9,000,000	7,200,000	1,590	1,113	722	185	104	66	47
9,500,000	7,600,000	1,679	1,175	762	195	110	69	49
10,000,000	8,000,000	1,767	1,237	802	206	116	73	52

Steam Velocity @ 10 PSIG								
Steam Pressure	10 PSIG	Pipe Size						
		2	2 1/2	4	6	8	10	12
Boiler Input	Boiler Output	**Feet Per Second Shaded cells are greater than 40 feet per second**						
200,000	160,000	31	22	14	4	2	1	1
300,000	240,000	47	33	21	5	3	2	1
400,000	320,000	63	44	29	7	4	3	2
500,000	400,000	79	55	36	9	5	3	2
600,000	480,000	94	66	43	11	6	4	3
700,000	560,000	110	77	50	13	7	5	3
800,000	640,000	126	88	57	15	8	5	4
900,000	720,000	141	99	64	16	9	6	4
1,000,000	800,000	157	110	71	18	10	6	5
1,100,000	880,000	173	121	78	20	11	7	5
1,200,000	960,000	188	132	86	22	12	8	6
1,300,000	1,040,000	204	143	93	24	13	8	6
1,400,000	1,120,000	220	154	100	26	14	9	6
1,500,000	1,200,000	236	165	107	27	15	10	7
1,600,000	1,280,000	251	176	114	29	17	10	7
1,700,000	1,360,000	267	187	121	31	18	11	8
1,800,000	1,440,000	283	198	128	33	19	12	8
1,900,000	1,520,000	298	209	135	35	20	12	9
2,000,000	1,600,000	314	220	143	37	21	13	9
2,500,000	2,000,000	393	275	178	46	26	16	11
3,000,000	2,400,000	471	330	214	55	31	19	14
3,500,000	2,800,000	550	385	250	64	36	23	16
4,000,000	3,200,000	628	440	285	73	41	26	18
4,500,000	3,600,000	707	495	321	82	46	29	21
5,000,000	4,000,000	785	550	357	91	52	32	23
5,500,000	4,400,000	864	605	392	101	57	36	25
6,000,000	4,800,000	942	660	428	110	62	39	28
6,500,000	5,200,000	1,021	715	464	119	67	42	30
7,000,000	5,600,000	1,099	770	499	128	72	45	32
7,500,000	6,000,000	1,178	825	535	137	77	49	34
8,000,000	6,400,000	1,256	880	571	146	83	52	37
8,500,000	6,800,000	1,335	935	606	155	88	55	39
9,000,000	7,200,000	1,414	989	642	164	93	58	41
9,500,000	7,600,000	1,492	1,044	677	174	98	61	44
10,000,000	8,000,000	1,571	1,099	713	183	103	65	46

Steam Velocity @ 12 PSIG								
Steam Pressure	12 PSIG	Pipe Size						
		2	2 1/2	4	6	8	10	12
Boiler Input	Boiler Output	**Feet Per Second Shaded cells are greater than 40 feet per second**						
200,000	160,000	37	19	7	3	2	1	1
300,000	240,000	41	29	11	5	3	2	1
400,000	320,000	55	39	15	6	4	2	2
500,000	400,000	69	48	18	8	5	3	2
600,000	480,000	82	58	22	10	5	3	2
700,000	560,000	96	68	25	11	6	4	3
800,000	640,000	110	77	29	13	7	5	3
900,000	720,000	124	87	33	14	8	5	4
1,000,000	800,000	137	97	36	16	9	6	4
1,100,000	880,000	151	106	40	18	10	6	4
1,200,000	960,000	165	116	44	19	11	7	5
1,300,000	1,040,000	179	126	47	21	12	7	5
1,400,000	1,120,000	192	135	51	22	13	8	6
1,500,000	1,200,000	206	145	54	24	14	8	6
1,600,000	1,280,000	220	155	58	26	14	9	6
1,700,000	1,360,000	234	164	62	27	15	10	7
1,800,000	1,440,000	247	174	65	29	16	10	7
1,900,000	1,520,000	261	184	69	30	17	11	8
2,000,000	1,600,000	275	193	73	32	18	11	8
2,500,000	2,000,000	344	242	91	40	23	14	10
3,000,000	2,400,000	412	290	109	48	27	17	12
3,500,000	2,800,000	481	338	127	56	32	20	14
4,000,000	3,200,000	550	386	145	64	36	23	16
4,500,000	3,600,000	618	435	163	72	41	25	18
5,000,000	4,000,000	687	483	181	80	45	28	20
5,500,000	4,400,000	756	531	200	88	50	31	22
6,000,000	4,800,000	825	580	218	96	54	34	24
6,500,000	5,200,000	893	628	236	104	59	37	26
7,000,000	5,600,000	962	676	254	112	63	40	28
7,500,000	6,000,000	1,031	725	272	120	68	42	30
8,000,000	6,400,000	1,099	773	290	128	72	45	32
8,500,000	6,800,000	1,168	821	308	136	77	48	34
9,000,000	7,200,000	1,237	869	326	144	81	51	36
9,500,000	7,600,000	1,306	918	345	152	86	54	38
10,000,000	8,000,000	1,374	966	363	160	90	57	40

Cleaning the Boiler

Steel or Cast Iron Hydronic Boilers

Consult with the manufacturer about the products they recommend. In most instances, Tri Sodium Phosphate or TSP is mixed with the water and circulated for 2-3 hours. Drain, flush and refill with fresh water. After flushing the cleaner from the system, a pH test should be done to verify that the water is within the proper range. The typical pH range is 7.0 - 8.5. Please check with the manufacturer to verify their requirements. Some cast iron boilers have special requirements due to the material that is used between the sections.

Note: If you have an aluminum water boiler, you may not be able to use Tri Sodium Phosphate (TSP) as a cleaning agent for the boiler. TSP has a high pH level and could remove the natural protective oxide layer from the aluminum.

Steel or Cast Iron Steam Boilers

When cleaning a steam boiler, do not allow the TSP to boil off into the piping. A typical application is to heat the boiler to 180 degrees F and allow it to stay there for 2 hours. After, drain, flush and refill the system. Check the pH level of the water to make sure it is within the manufacturers recommended limits.

I like to let the condensate run to the drain for a couple days when it is started to allow the system to purge itself of the contaminants.

Glycol for the System

If you are contemplating the use of glycol in your hydronic system, there are several important factors to consider.

Can Your Boiler Tolerate Glycol? Some boilers, such as aluminum ones, do not react well with standard glycol. Before installing glycol in your system, check with the boiler manufacturer. Glycol designed for aluminum boilers can be ordered. Some boiler manufacturers have specific requirements when adding glycol for their boilers.

Reduced Efficiency When adding glycol to a hydronic heating system, it will reduce the heat transfer efficiency of the system. For example, when you add 20% propylene glycol to the system, the heating system efficiency drops 3% lower. At 50% concentration, your capacity drops 10%. The 80% boiler suddenly dropped to a little over 70% efficient. This should be factored in when sizing the new system. In addition, the compression tank will have to be oversized by 20% to compensate for the glycol. The pumps also lose about 10% efficiency when the system is filled with a 50% mixture of glycol and water.

Increased Maintenance Systems containing glycol require extra maintenance. The pH has to be checked yearly. The glycol composition has to be checked using a refractometer. Glycol will breakdown eventually and may cause damage to your system. If the concentration level drops in your tests, this should indicate a piping leak. The leak should be found and repaired.

Cleaning the system The system should be cleaned prior to introduction of glycol. Flush the system with a heated 1-2% solution of Trisodium phosphate for 2 to 4 hours, then drain and rinse thoroughly. This will remove excess pipe dope, cutting oils and solder flux.

Automotive Glycol? This is not recommended as it was not designed to be used in a hydronic system. Automotive antifreeze is formulated with silicates, which tend to gel, reducing heat transfer efficiency. Use an inhibited glycol designed for heat transfer applications.

Which Type of Glycol? You can choose between ethylene and inhibited propylene glycol. Uninhibited glycol is very corrosive and could lead to damage in your system. Ethylene glycol is toxic to humans and animals. A special permit may be required when using it. Ethylene glycol will provide slightly better freeze protection than propylene glycol. Propylene glycol is more environmentally friendly and not toxic.

How Much Glycol? The most common percentage is between 20% to 50% concentration of glycol.

Water Quality Water chemistry is a concern when you introduce glycol into your system. Poor water quality can lead to scale, sediment deposits, or the creation of sludge in the heat exchanger which will reduce heat transfer efficiency. It can damage the system by depleting the corrosion inhibitor and promoting a number of corrosions including general and acidic attack corrosion. Before using the local water, it should be analyzed by an water treatment expert. Good quality water contains:

- Less than 50 ppm of calcium
- Less than 50 ppm of magnesium
- Less than 100 ppm (5 grains) of total hardness
- Less than 25 ppm of chloride
- Less than 25 ppm of sulfate

The glycol should be mixed at room temperature with either demineralized or deionized water if the water quality is questionable.

The safest, although not the least expensive, solution may be to order premixed glycol from the manufacturer.

Freezing Point

Concentration by volume	Ethylene Glycol	Propylene Glycol
50%	-37F	-28F
40%	-14F	-13F
30%	+2F	+4F
20%	+15F	+17F

Maintenance The glycol must be checked at least once a year in accordance with the manufacturer's recommendations. A base line analysis should be performed within two to four weeks of initial mixing. This measurement will be used to verify that the fill was completed properly, and will serve as a reference point for comparison with future test results. As a bare minimum, the solution should be analyzed for glycol concentration, solution pH and general fluid quality.

Solution Testing If you are using a 30% to 50% solution, the pH should be between 8.3 and 9.0. If the level falls below 8.0, this could indicate lowered inhibitors. In some instances, inhibitors can be added. If the pH falls below 7.0, the glycol should be removed and flushed. This level of pH could cause damage to the boiler and piping. The system should be tested once a month.

Mark the System When you are done with the system and turn it over to the owner, the system should have clear signage that tells anyone working on the system that it contains glycol. It should also have the following information:

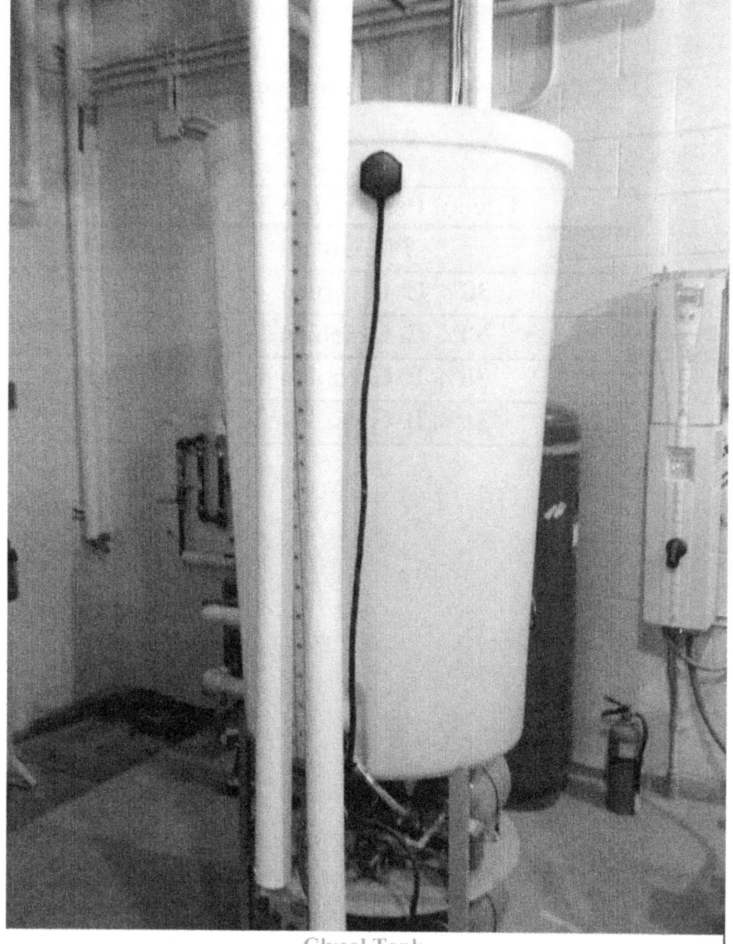

Glycol Tank

- Type of glycol that is used in the system, ethylene or propylene
- Concentration of glycol
- Initial pH readings
- Installation date of glycol
- Last test date

Some Glycol Considerations Galvanized pipe should not be used in the hydronic system as the Zinc could have an adverse effect and form sludge.

Make sure the system is clean before filling. Pre-fill flushing is highly recommended.

Mix the solution at room temperature.

In order to minimize the possibility of glycol loss due to undetected leaks, hydrostatically test the system for 24 hours prior to filling.

Never use a chromate water treatment in a system with glycol. The chromate will damage the glycol and can lead to severe system degradation.

Do not use in a system that may have a solution temperature over 300F.

Do not use check valves or closed zone valves that would isolate a part of the system, preventing proper expansion and resulting in freeze damage.

A strainer, sediment trap, or some other means for cleaning the piping system must be provided. It should be located in the return line ahead of the boiler and pump. This must be cleaned frequently during initial operation.

Automatic make-up water systems should be avoided in order to prevent undetected dilution or loss of glycol.

Check local codes to see if systems containing these solutions must include a back-flow preventer, or an actual disconnect from city water lines.

Do not use glycol in steam systems.

BTU per Hour with Glycol

Glycol Percentage & Type	Formula
No Glycol	BTUH = GPM x 500 x Δt ^{0}F
30% E. Glycol @ 68 ^{0}F	BTUH = GPM x 445 x Δt ^{0}F
50% E. Glycol @ 32 ^{0}F	BTUH = GPM x 395 x Δt ^{0}F
30% P. Glycol @ 68 ^{0}F	BTUH = GPM x 465 x Δt ^{0}F
50% P. Glycol @ 32 ^{0}F	BTUH = GPM x 420 x Δt ^{0}F

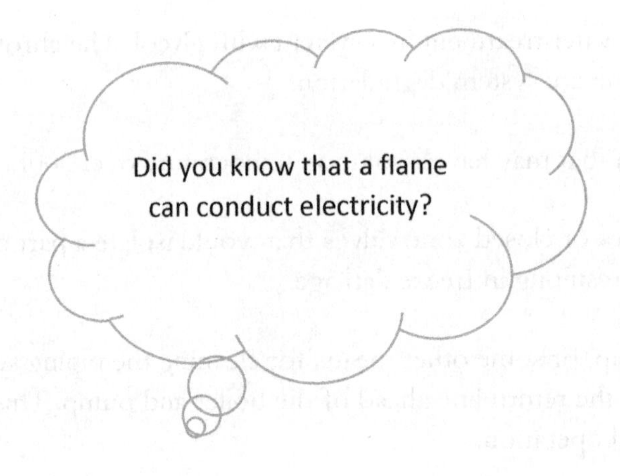

Trouble Shooting Boilers with Atmospheric Burner

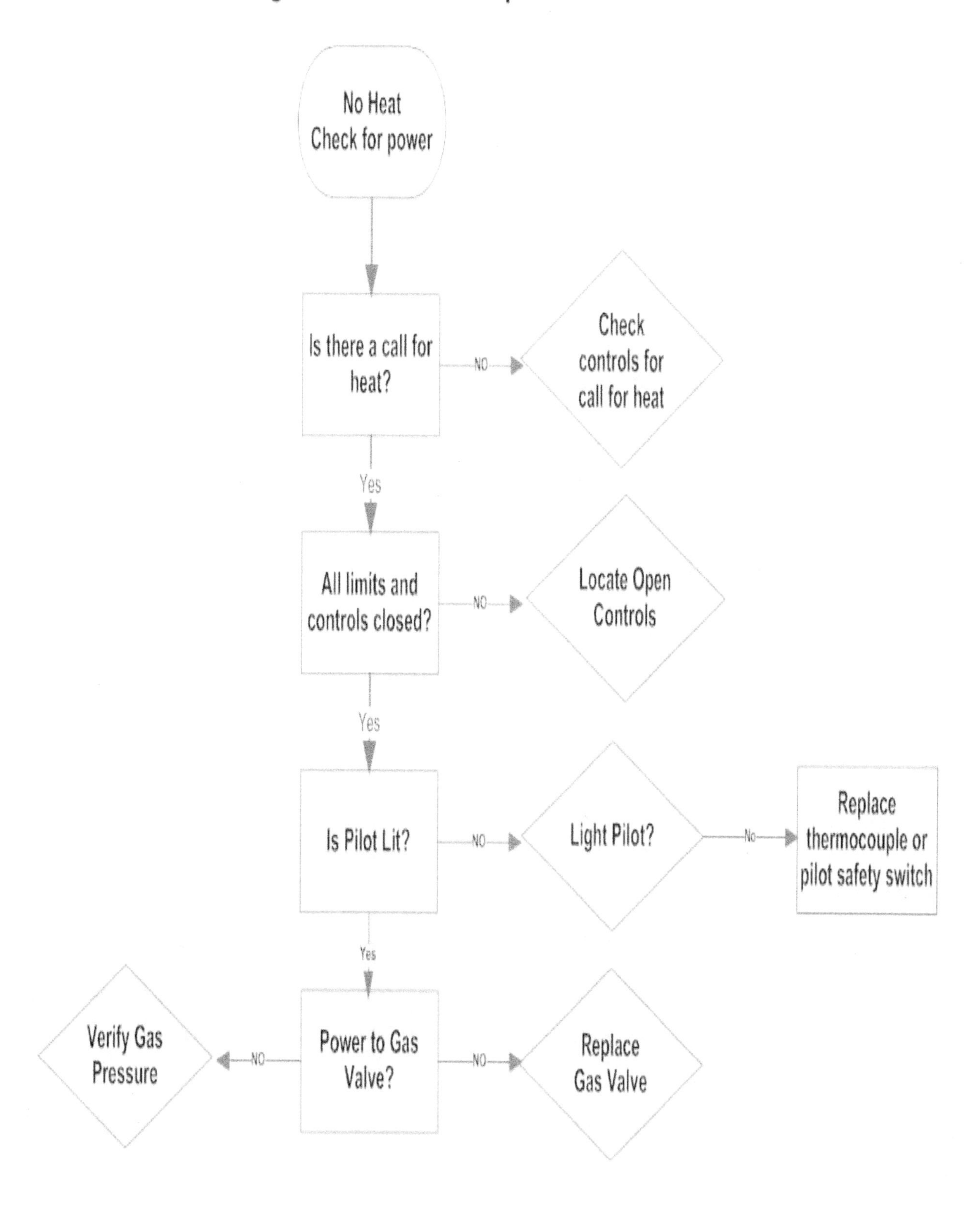

Trouble Shooting Boilers with Power Burner

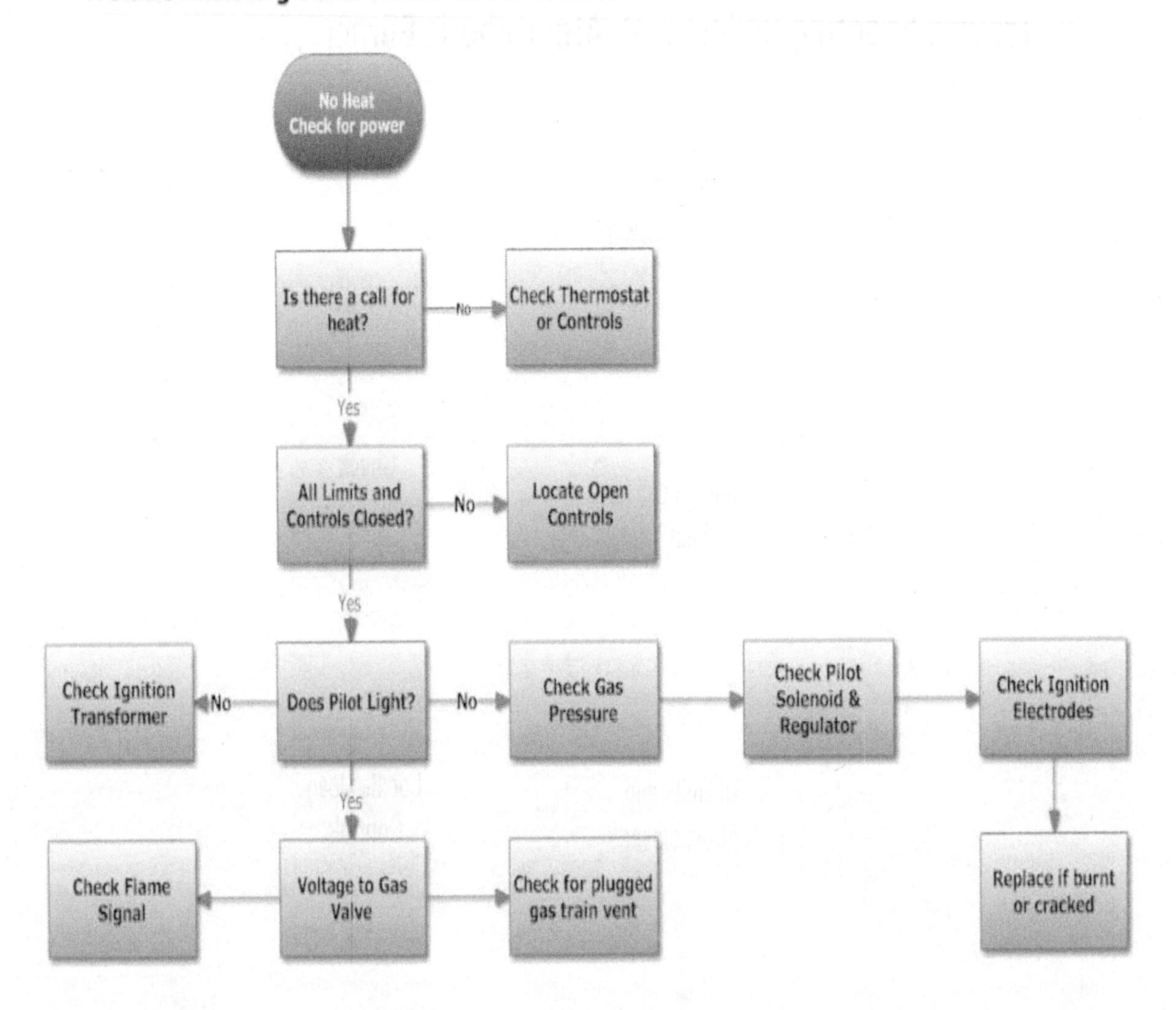

Possible Manual Reset Controls
Gas Pressure Switches
Low Water Cutout
Flame Safeguard
High Limit Controls
Motor Starter

Clocking a Gas Meter

Many engineers specify that the new boilers be "clocked" to assure proper firing rate. Clocking a gas meter is a way of verifying the actual firing rate of the boiler. There are a couple of ways to perform that task. To properly "clock" a meter, be sure that your boiler is the only apparatus that is firing. The boiler should also be at high fire. A way to check this is to shut off all the items in the boiler room and observe the gas meter. If the hands on the dial are still moving, there are other items consuming gas. It is sometimes difficult to perform this task in a commercial building, as there are could be many items operating simultaneously.

Some commercial gas meters require you to compensate for the temperature as well as the pressure of the gas. When the meter was calibrated, it was at a certain temperature and gas pressure at the factory. The meter above was calibrated using 60 degree F gas. The field conditions may be different. In the end of the chapter, there are some figures that will help you to compensate for the conditions found at the job site.

How to Clock the Meter

Once you are sure that there are no other appliances using gas, you will need to start your unit and assure that it is firing at full rate. You then want to count the number of revolutions the most sensitive dial on the gas meter makes in one minute. Most natural gas has a heating value of 1,000 BTU/cubic foot. Let us assume that our most sensitive dial is ½ cubic feet per revolution.

A. Count the revolutions the ½ cubic foot dial makes in one minute.
B. Multiply the revolutions by 30,000 to obtain the firing rate in Btu's/ Hr

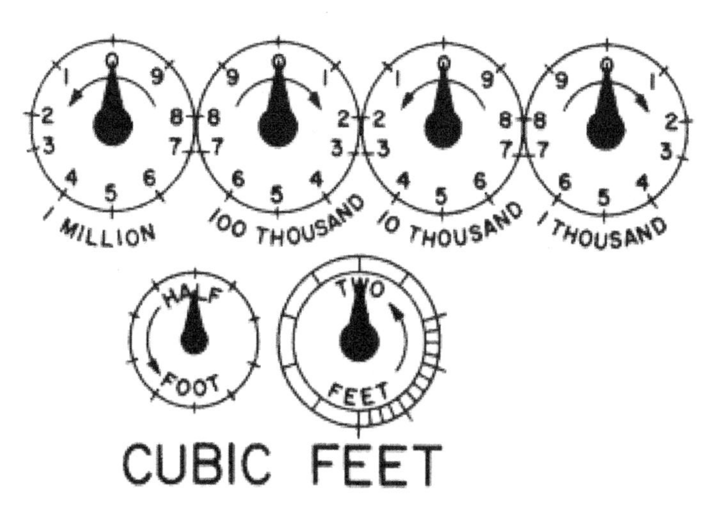

For example, the ½ cubic foot dial made 3.2 revolutions in one minute. The boiler is firing at 3.2 revolutions x 30,000 BTU/revolution = 96,000 BTUH. If you find that the heating value is different from 1,000 Btu's per cubic foot, you would have to make an adjustment. The local gas company could inform you of the heating content of their gas. For example, if the gas company tells you that the heating value is 1,050 BTU/ cubic foot, you would need to adjust your final reading. 1,050 BTU/ Cubic foot (Actual BTU) divided by 1,000 Btu/ Cubic foot (This was assumed to be the BTU content) = 1.05. Therefore, to recalculate the new rate, we would multiply 96,000 Btuh (From above) x 1.050 = 100,800 Btu/ HR. This is the actual firing rate of the appliance.

The 30,000 calculation only works with ½ cubic foot dial. For other size dials, see below.

Remember our basic formula is Number of revolutions x factor below = BTU/ Hr. This is based on 1,000 BTU/ Cubic Foot.

NOTE: To get a more accurate reading, it is better to allow the test to be done for a longer time. I would recommend 5 minutes. You would then divide the reading by 5 to get the average. The chart below features different timing for the dials, up to five minutes. For example, if the 5 "Cubic Feet per Revolution" dial made two revolutions in five minutes, your firing rate would be as follows:

60,000 x 2 = 120,000 Btuh

Multiplying Factor for Gas Meter

Cubic Feet per Revolution	1 Minute Timing	2 Minute Timing	3 Minute Timing	5 Minute Timing
	BTUH			
½	30,000	15,000	7,500	6,000
1	60,000	30,000	15,000	12,000
2	120,000	60,000	30,000	24,000
5	300,000	150,000	75,000	60,000
Based on 1,000 Btu per cubic foot of gas				

Clocking a Gas Meter Option 2

A second method for "clocking" a gas meter is as follows:

Start the boiler; making certain that no other gas-fired appliance is operating. Measure the amount of time it takes for the smallest dial to make one complete revolution. In the above dials, the ½ cubic foot dial is the timing dial.

Refer to a natural gas timing chart under ½ cubic foot column and see what the input is to your boiler.

Check and compare the calculated input with the input rating on the heating unit data plate. If the unit is under-fired or over-fired by more than 10%, check the gas pressure to the unit with a fluid filled manometer and adjust as necessary.

(For example, the unit being tested takes 29 seconds for the ½ cubic foot dial to make one complete revolution. Using the chart, this translates to 62 cubic feet per hour. Based upon the assumption that one cubic foot of natural gas has 1,000 BTU's (Check with your local utility for actual BTU content), the calculated input is 62,000 BTU's per hour.

You will get a better reading by allowing the dial to rotate several times and dividing the total by the amount of revolutions to get an average. In the above example, if it took 1 minute, 27 seconds or 87 seconds to make three revolutions, our average input would be 29 seconds.

Natural Gas Timing Chart in Cubic Feet/ Hour

Seconds for one revolution	1/2 Cu Ft	1 Cu Ft	2 Cu Ft	5 Cu Ft
10	180	360	720	1,800
11	164	327	655	1,636
12	150	300	600	1,500
13	138	277	555	1,385
14	129	257	514	1,285
15	120	240	480	1,200
16	112	225	450	1,125
17	106	212	424	1,059
18	100	200	400	1,000
19	95	189	379	947
20	90	180	360	900
21	86	171	345	857
22	82	164	327	818
23	78	157	313	783
24	75	150	300	750
25	72	144	288	720
26	69	138	277	692
27	67	133	267	667
28	64	129	257	643
29	62	124	248	621
30	60	120	240	600
31	58	116	232	581
32	56	113	225	563
33	55	109	218	545
34	53	106	212	529
35	51	103	205	514
36	50	100	200	500
37	49	97	195	486
38	47	95	189	474
39	46	92	185	462
40	45	90	180	450
40	45	90	180	450

Seconds for one revolution	1/2 Cu Ft	One Cu Ft	Two Cu Ft	Five Cu Ft
41	44	88	176	440
42	43	86	172	430
43	42	84	167	420
44	41	82	164	410
45	40	80	160	400
46	39	78	157	391
47	38	77	153	383
48	37	75	150	375
49	37	73	147	367
50	36	72	144	360
51	35	71	141	353
52	35	69	138	346
53	34	68	136	340
54	33	67	133	333
55	33	65	131	327
56	32	64	129	321
57	32	63	126	316
58	31	62	124	310
59	30	61	122	305
60	30	60	120	300
62	29	58	116	290
64	29	56	112	281
66	29	54	109	273
68	28	53	106	265
70	26	51	103	257
72	25	50	100	250
74	24	48	97	243
76	24	47	95	237
78	23	46	92	231
80	22	45	90	225
82	22	44	88	220
84	21	43	86	214
86	21	42	84	209
88	20	41	82	205

NOTES:

On a commercial gas meter, you may have to calculate a pressure and/or temperature correction factor. You will need to contact the local gas company for this factor.

Gas Pressure Correction Factor

Actual Meter Pressure (psi)			Meter Base Pressure		
	4 oz or 7" w.c.	8oz or 14" w.c.	10 oz or 17.5" w.c.	1 Psi or 28" w.c.	2 psi or 56" w.c.
0	0.983	0.966	0.958	0.935	0.878
¼	1.0	0.983	0.975	0.951	0.893
½	1.017	1	0.992	0.968	0.909
5/8	1.026	1.008	1	0.976	0.916
1	1.051	1.034	1.025	1.000	0.929
2	1.119	1.101	1.092	1.065	1.000
3	1.188	1.168	1.158	1.130	1.061
4	1.256	1.235	1.225	1.195	1.122
5	1.324	1.302	1.291	1.260	1.183
6	1.392	1.369	1.358	1.325	1.244
7	1.461	1.436	1.424	1.390	1.305
8	1.529	1.503	1.491	1.455	1.366
9	1.597	1.570	1.557	1.520	1.427
10	1.666	1.638	1.624	1.584	1.488

Gas Temperature Correction Factor

Gas Temperature Degrees F	Meter Calibration Temperature				
	60 Degrees F	65 Degrees F	68 Degrees F	70 Degrees F	72 Degrees F
0	1.130	1.141	1.148	1.152	1.157
5	1.118	1.129	1.135	1.140	1.144
10	1.106	1.117	1.123	1.128	1.132
15	1.095	1.105	1.112	1.116	1.120
20	1.083	1.094	1.100	1.104	1.108
25	1.072	1.082	1.089	1.093	1.097
30	1.061	1.071	1.078	1.082	1.086
35	1.051	1.061	1.067	1.071	1.075
40	1.040	1.050	1.056	1.060	1.064
45	1.030	1.040	1.046	1.050	1.053
50	1.020	1.029	1.035	1.039	1.043
55	1.010	1.019	1.025	1.029	1.033
60	1.000	1.010	1.015	1.019	1.023
65	0.990	1.000	1.006	1.010	1.013
70	0.981	0.991	0.996	1.000	1.004
75	0.972	0.981	0.987	0.991	0.994
80	0.963	0.972	0.978	0.981	0.985
85	0.954	0.963	0.969	0.972	0.976
90	0.945	0.955	0.960	0.964	0.967
95	0.937	0.946	0.951	0.955	0.959
100	0.929	0.938	0.943	0.946	0.950

Radiator Ratings

Hydronic and Steam

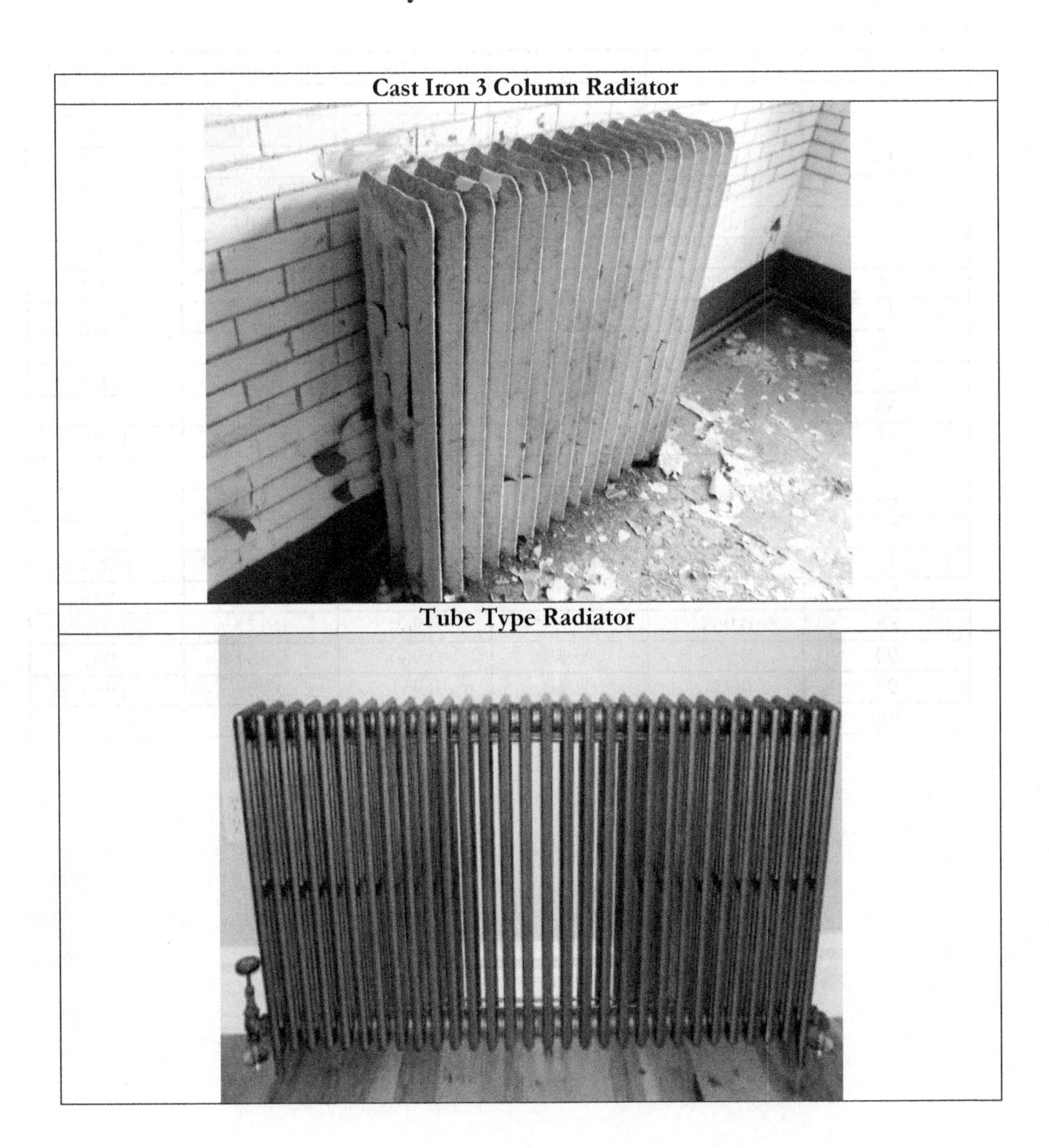

Cast Iron 3 Column Radiator

Tube Type Radiator

Hydronic

Estimated Btu's per section of hydronic radiator *Based on 180 degree supply water & 70 degree F room temperature.*									
Radiator Height	Number of Tubes					*Number of Columns*			
	3	4	5	6	7	*1*	*2*	*3*	*4*
13"					447				
16 1/2"					595				
18"								*383*	*510*
20"	298	383	452	510	723	*255*	*340*		
22"								*510*	*680*
23"	340	425	510	595		*282*	*396*		
26"	359	468	595	680		*340*	*452*	*638*	*850*
30"	510								
32"		595	735	850		*425*	*566*	*765*	*1,105*
36"	595								
37"		701	850	1,020					
38"						*510*	*680*	*850*	*1,360*
45"							*850*	*1,020*	*1,700*

Low Pressure Steam

Estimated Btu's per section of steam radiator *Based on 2 psi steam & 70 degree F room temperature.*									
Radiator Height	Number of Tubes					*Number of Columns*			
	3	4	5	6	7	*1*	*2*	*3*	*4*
13"					630				
16 1/2"					840				
18"								*540*	*720*
20"	420	540	638	720	1,020	*360*	*480*		
22"								*720*	*960*
23"	480	600	720	840		*398*	*559*		
26"	506	660	840	960		*480*	*638*	*900*	*1,200*
30"	720								
32"		840	1,039	1,200		*600*	*799*	*1,080*	*1,560*
36"	840								
37"		990	1,200	1,440					
38"						*720*	*960*	*1,200*	*1,920*
45"							*1,200*	*1,440*	*2,400*

Radiator Size Chart

1 Column	2 Column	3 Column
	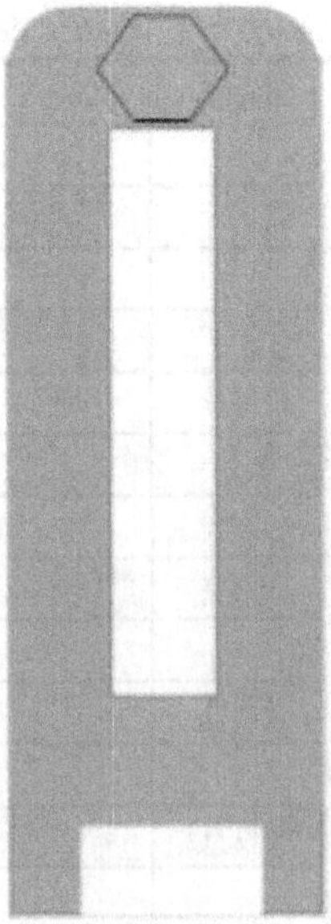	
4 Column	**5 Column**	**6 Column**

Cast Iron Radiator Ratings

Column Radiators

Column Radiators One Column			
Height	Sq Ft / Section	Hydronic 180 degree F Btuh per Section	Steam 2 Psig Btuh per Section
20”	1.5	255	360
23”	1.66	282	398
26”	2	340	480
32”	2.5	425	600
38”	3	510	72

Column Radiators Two Column			
Height	Sq Ft / Section	Hydronic 180 degree F Btuh per Section	Steam 2 psig Btuh per Section
20”	2	340	480
23”	2.33	396	559
26”	2.66	452	638
32”	3.33	566	799
38”	4	680	960
45”	5	850	1,200

Column Radiators Three Column			
Height	Sq Ft / Section	Hydronic 180 degree F Btuh per Section	Steam 2 psig Btuh per Section
18”	2.25	383	540
22”	3	510	720
26”	3.75	638	900
32”	4.5	765	1,080
38”	5	850	1,200
45”	6	1,020	1,440

Column Radiators Four Column			
Height	Sq Ft / Section	Hydronic 180 degree F Btuh per Section	Steam 2 psig Btuh per Section
18”	3	510	720
22”	4	680	960
26”	5	850	1,200
32”	6.5	1,105	1,560
38”	8	1,360	1,920
45”	10	1,700	2,400

Cast Iron Radiator Ratings
Thin Tube Radiators

Three Tube			
Height "	Sq. Ft. per Section	Hydronic 180 degree F Btuh per Section	Steam 2 psig Btuh per Section
20"	1.75	298	420
23"	2	340	480
26"	2.11	359	506
30"	3	510	720
36"	3.5	595	840

Four Tube			
Height "	Sq. Ft. per Section	Hydronic 180 degree F Btuh per Section	Steam 2 psig Btuh per Section
20"	2.25	383	540
23"	2.5	425	600
26"	2.75	468	660
32"	3.5	595	840
37"	4.125	701	990

Five Tube			
Height "	Sq. Ft. per Section	Hydronic 180 degree F Btuh per Section	Steam 2 psig Btuh per Section
20"	2.66	452	638
23"	3	510	720
26"	3.5	595	840
32"	4.33	735	1039
37"	5	850	

Six Tube			
Height "	Sq. Ft. per Section	Hydronic 180 degree F Btuh per Section	Steam 2 psig Btuh per Section
20"	3	510	720
23"	3.5	595	840
26"	4	680	960
32"	5	850	1,200
37"	6	1,020	1,440

Seven Tube			
Height "	Sq. Ft. per Section	Hydronic 180 degree F Btuh per Section	Steam 2 psig Btuh per Section
13"	2.625	447	630
16 1/2"	3.5	595	840
20"	4.25	723	1,020

Did you know that a steam radiator only heats all the way across on the coldest days of the year?

Industry Links that I like

American Boiler Manufacturers Association	www.abma.com
American Gas Association	www.aga.org
American Society of Mechanical Engineers	www.asme.org
Appropriate Designs John Siegenthaler	www.hydronicpros.com
ASHRAE	www.ashrae.org
Engineered Systems Magazine	www.esmagazine.com
Fire & Ice *(My company)*	www.fireiceheat.com
Foley Mechanical	www.foleymechanical.com
FW Behler Dave Yates	www.fwbehler.com
Healthy Heating Robert Bean	www.healthyheating.com
Heating Help Dan & Erin Holohan	www.heatinghelp.com
Heat Tech	www.heattech.com
Johnson Controls	www.johnsoncontrols.com
Mechanical Hub	www.mechanical-hub.com
National Board of Boiler and Pressure Vessel Inspectors	www.nationalboard.org
Oil & Energy Service Professionals OESP	www.thinkoesp.org
Plumbing & Mechanical Magazine	www.pmmag.com
PM Engineer Magazine	www.pmengineer.com
Radiant Professionals Alliance	www.radiantprofessionalsalliance.org
Rel-Tek	www.rel-tek.com
Rite Boiler	www.riteboiler.com
RSES	www.rses.org
Siemens	http://w3.usa.siemens.com/buildingtechnologies/us
TF Campbell	www.tfcampbell.com
Triad Boiler Systems	www.triadboiler.com

Heating Formulas and Rules of Thumb Index

Heating Formulas & Rules of Thumb

All Boilers

These are rules of thumb. Please consult with the equipment manufacturer for actual requirements.

Combustion Air

Combustion Air Openings

Each fuel-burning piece of equipment requires combustion air to operate safely. The following are some guidelines to help you see whether the existing combustion air louvers will be adequate for the replacement project if you are directly venting the flues.

Number of openings required = 2

Each boiler room should have two openings. One should be within one foot of ceiling and the other opening within one foot of the floor. This does not include boiler rooms with direct vented appliances.

Combustion Air from Outside

Direct Openings to Outside	1" Free area for each 4,000 Btuh
Horizontal Openings to Outside	1" Free space for each 2,000 Btuh
Vertical Openings to Outside	1" Free space for each 4,000 Btuh
Mechanical Ventilation from Outside	1 cfm per 2,400 Btuh

Indoor Combustion Air 50 cubic feet of volume for each 1,000 Btuh input of all the appliances.

Ducted Combustion Air to Burner- Power Flame recommends sizing the duct for a pressure drop of 0.1" w.c. including all screens, filters, and fittings.

This chart shows the CFM required for mechanical boiler room ventilation using fans. CFM Required @ Various Boiler Btuh			
Btuh	CFM	**Btuh**	CFM
50,000	21	**1,100,000**	458
75,000	31	**1,500,000**	625
100,000	42	**2,000,000**	833
200,000	83	**2,500,000**	1,042
300,000	125	**3,000,000**	1,250
500,000	208	**3,500,000**	1,458
700,000	292	**4,000,000**	1,667
900,000	375	**4,500,000**	1,875

Direct Vent Combustion Air Sizing

If you are directly venting combustion air for the boiler, this may help you when sizing the air duct. The chart below is based on using 15 cubic feet of air for every cubic foot of gas burned or about 50% excess air. Most new boilers will use about 12 parts air or 20% excess air. Verify that with the boiler manfacturer.

Estimated Combustion Air Required @ Various Boiler Inputs			
Boiler Btuh Input	Cu Ft Gas / Hr**	Cu Ft Gas/ Minute	**CFM Air***
200,000	200	3.33	**50**
300,000	300	5.0	**75**
400,000	400	6.67	**100**
500,000	500	8.33	**125**
600,000	600	10.0	**150**
700,000	700	11.67	**175**
800,000	800	13.33	**200**
900,000	900	15.0	**225**
1,000,000	1,000	16.67	**250**
1,500,000	1,500	25.0	**375**
2,000,000	2,000	33.33	**500**
3,000,000	3,000	50.0	**750**
4,000,000	4,000	66.67	**1,000**
5,000,000	5,000	83.33	**1,250**
6,000,000	6,000	100.0	**1,500**
7,000,000	7,000	116.7	**1,750**
8,000,000	8,000	133.3	**2,000**
9,000,000	9,000	150.0	**2,250**
10,000,000	10,000	166.7	**2,500**
*** Based upon 15 cubic feet of air for every cubic foot of gas burned.**			
**** Based on 1,000 Btu per cubic foot of gas**			

How Excess Combustion Air Affects Boiler Condensing Temperatures			
O2%	CO2%	Excess Air %	Dew Point ^{0}F
3.0%	10.0%	15.0%	133^{0}F
4.0%	9.50%	20%	131^{0}F
5.0%	9.0%	29.0%	130^{0}F
6.0%	8.40%	36.0%	128^{0}F
7.0%	7.9%	46.5%	123^{0}F
8.0%	7.3%	56.5%	122^{0}F
9.0%	6.7%	68.6%	118^{0}F
10.0%	6.2%	83.5%	116^{0}F
11.0%	5.6%	100%	113^{0}F

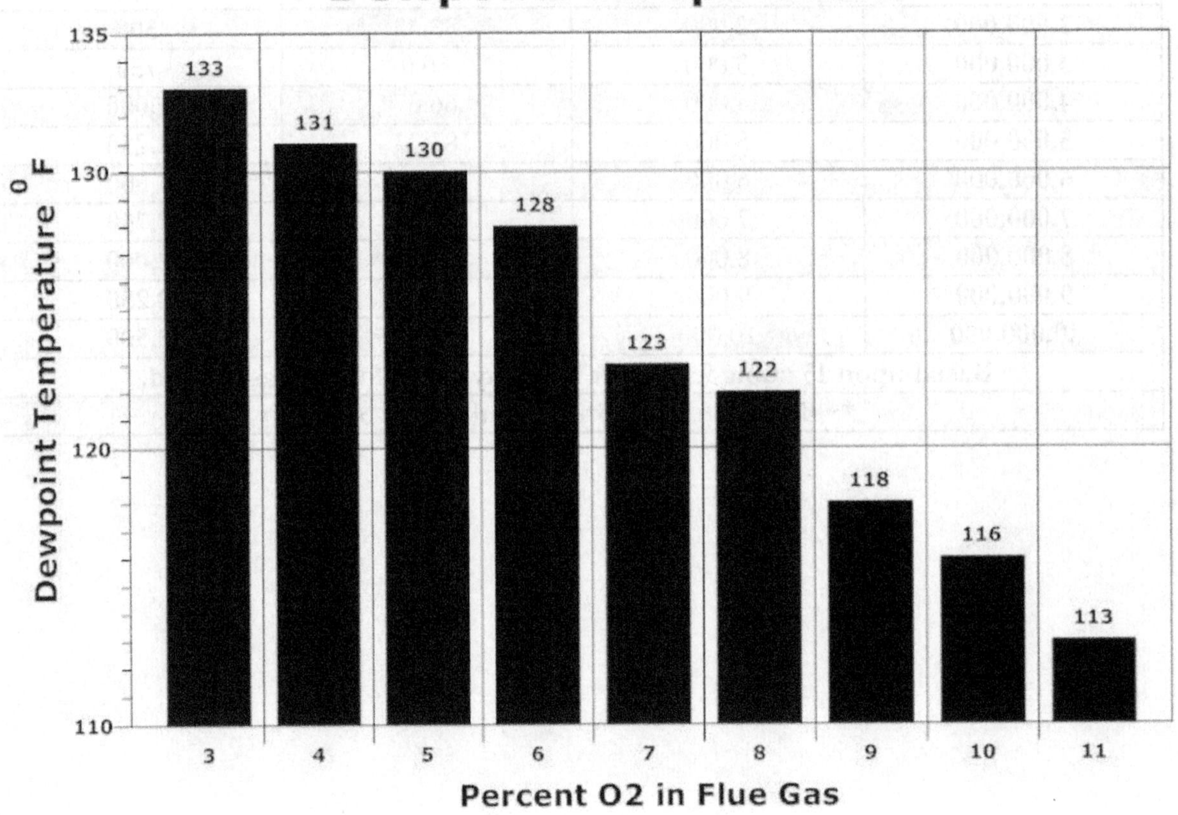

Venting the Gas Train

Sizing Gas Train Manifold Vent
Source: Philadelphia Gas Works

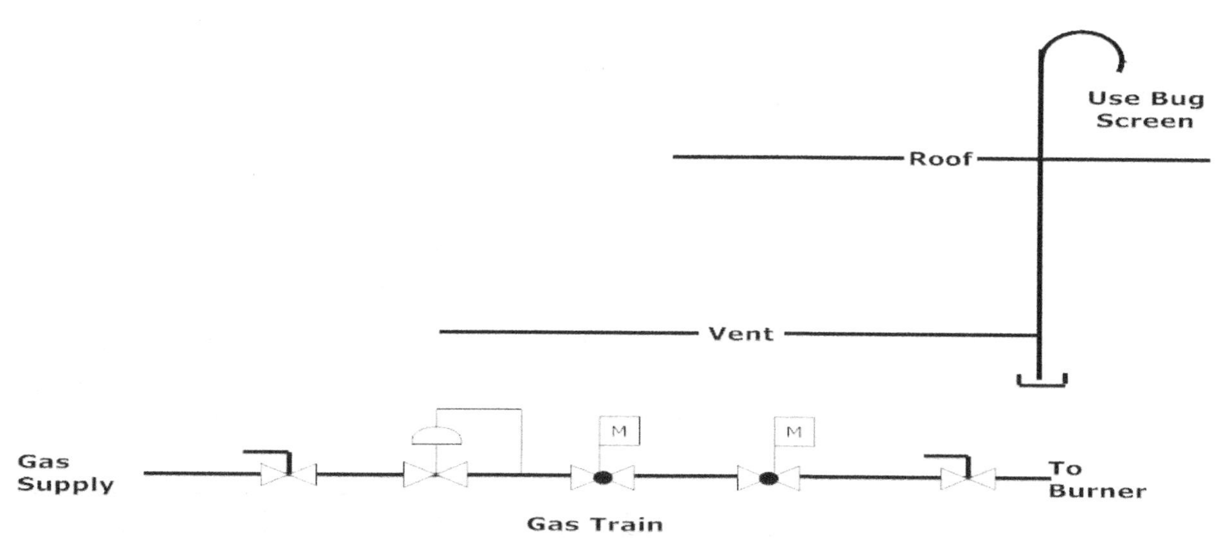

Some gas train components, such as regulators and gas pressure switches, have to be vented to the outside. A leaking 1/4" fitting could lose enough gas to fill a 10 foot by 10 foot x 10 foot high room with a combustible mixture within an hour. When combining common vents from gas train components, you need to see if the common pipe cross sectional area is large enough for all the components.

Let us assume that we will be venting two regulators with 3/4" vents and two gas pressure switches with 1/4" vents. We will get our cross sectional sizes from the chart below

3/4" regulators	0.533 X 2 = 1.066
1/4" gas pressures switches	0.104 X 2 = 0.208
	Total 1.274

We would need a pipe size of 1 1/4" to combine all these vents.

- Pipe runs over 30 feet horizontal are not recommended and should be increased one pipe size for each 30 feet run.
- Normally open vent valves cannot be combined with other vents or other appliances.

Vent terminations should be:

- 4 feet below, 1 foot above, and 3 feet horizontally from windows, doors, and gravity air intakes. 3 feet above any forced air inlet within 10 feet horizontally.
- Vents should have a screened 90 facing down.

ASME CSD1 Sizing: If your locale follows the ASME CSD1 Code, Section CF-190 uses a different formula for sizing the manifolded vent line. It requires that *"the manifolded line shall have a cross-sectional area not less than the area of the largest branch line directly piped to the manifolded line plus 50% of the additional cross sectional areas of the manifolded branch lines."*

Pipe Size	Inside Diameter	Inside Cross Sectional Area	50% of Inside Cross Sectional Area
1/8"	0.269	**0.057**	**0.03**
1/4"	0.364	**0.104**	**0.05**
1/2"	0.622	**0.304**	**0.15**
3/4"	0.824	**0.533**	**0.27**
1"	1.049	**0.864**	**0.43**
1 1/4"	1.38	**1.495**	**0.75**
1 1/2"	1.61	**2.036**	**1.02**
2"	2.07	**3.36**	**1.68**
2 1/2"	2.469	**4.788**	**2.39**
3"	3.068	**7.393**	**3.70**
4"	4.026	**12.73**	**6.37**
5"	5.047	**20.004**	**10.02**
6"	6.065	**28.89**	**14.45**

If you would like to combine the same size vents, this table can help.

Combining Gas Train Vents using Full Area of Vent Pipes					
	Number of Vents				
Vent Pipe Size	2	3	4	5	6
¼"	**½"**	**¾"**	**¾"**	**¾"**	**1"**
½"	**1"**	**1 ¼"**	**1 ¼"**	**1 1/2"**	**1 ½"**
¾"	**1 ¼"**	**1 ½"**	**2"**	**2"**	**2"**

Combined vents using 50% Area

To meet ASME CSD1, you would have to use the full vent of the largest vent line and 50% of the all the smaller ones.

Combining Gas Train Vents using 50% Area of Vent Pipes					
	Number of Vents				
Vent Pipe Size	2	3	4	5	6
¼"	**0.1"**	**0.15"**	**0.2"**	**0.25"**	**0.3"**
½"	**0.3"**	**0.45"**	**0.6"**	**0.75"**	**0.9"**
¾"	**0.54"**	**0.81"**	**1.08"**	**1.35"**	**1.62"**

How Long Will It Last?
Source ASHRAE

Equipment	Years
Boilers	24-35
Burners	21
Boilers, Condensing	10-15*
Rooftop Units	15
Furnaces	18
Pumps, Base Mounted	20
Pumps Pipe Mounted	10
Pumps, Condensate	15
Condensate Piping	10-20
Steam Traps	*7 based upon US GSA*
*** Based upon findings by Chartered Institute of Building Service Engineers**	

Boiler HP to Boiler Output Btuh			
Boiler HP	Btu/Hr Output	Boiler HP	Btu/Hr Output
20	669,500	**250**	8,368,750
30	1,004,250	**300**	10,042,500
40	1,339,000	**350**	11,716,250
50	1,673,750	**400**	13,390,000
60	2,008,500	**450**	15,063,750
70	2,343,250	**500**	16,737,500
80	2,678,000	**600**	20,085,000
100	3,347,500	**700**	23,432,500
125	4,184,375	**800**	26,780,000
150	5,021,250	**1,000**	33,475,000
200	6,695,000		

Misc. Boiler Information

1 Boiler HP =	**34,500 Btuh**
	34.5 Lbs Steam/ Hr from and at 212 degrees F
	34.5 Lb H2O/ Hr
	140 EDR
	about 10-11 ½ square feet of boiler heating surface
	0.069 GPM
	4.14 GPH
1 HP =	0.746 KW
	746 WATTS
	2,545 BTUH
	1.0KVA
1 EDR =	150 Btuh Water
	240 Btuh Steam
	0.000496 gpm
	0.25 lbs steam cond/hr
1,000 EDR =	0.496 GPM
1 Btu =	will raise 1 cubic feet of air 55 degrees
	will raise 55 cubic feet of air 1 degree F
	Amount of heat required to raise one pound of water one degree
1 Lb Steam =	0.002 GPM
1 Watt =	3,415 BTUH
1,000,000 Btu =	1 MCF Gas
	1,000 Cubic feet of natural gas
	1 dekatherm of natural gas
	10 therms of natural gas
	293.1 KW of electricity
	7.29 gallons of #2 fuel oil
	10.93 gallons of propane
	1,000 pounds of steam
	29.31 boiler horsepower

Common Heating Calculations	
To Find?	**Perform this calculation**
Lbs Steam per Hour	Boiler HP x 34.5
Evaporation Rate GPM	Boiler HP x .069
MBTU per Hr Output (MBH)	Boiler HP x 33.4
Sq. Ft of Equivalent Direct Radiation (EDR)	Boiler HP x 139
Sq Ft Equivalent Direct Radiation (EDR)	BTU/Hr /240
Evaporation Rate GPM	EDR/1000 x 0.5
Evaporation rate GPM	Lbs Steam Hr / 500
BTU	500 x GPM x Delta T
GPM	BTU / 500 x Delta T
Delta T	BTU /500 x GPM
Calculate Sensible Heat on Fan	1.08 x CFM x Temp Rise or Delta T
Calculate Sensible Heat	500 x GPM x Temp Rise or delta T
Average Building Heat Loss	25-40 BTUH per square foot

Convert Temperature Readings

Fahrenheit to Celsius	Celsius to Fahrenheit
$\frac{\text{Degrees F} - 32}{1.8}$	$(1.8 * \text{ Degrees C}) + 32$

Types of Heat Transfer

Conduction is the transfer of heat through a material or substance. It could even transfer to the adjoining material. An example of this is the heat that transfers from the frying pan on a stove to the handle of the pan. In our industry, it is like the heat that transfers along a pipe as it is soldered. Heat is also conducted through the ceilings, walls and floors of homes.

Convection is the transfer of heat by a liquid or gas (such as air). Circulatory air motion due to warmer air rising and cooler air falling is a common mechanism by which thermal energy is transferred. In our industry, it would be the heat that is circulated in a room heated with fin tube radiation. Convective heat loss also occurs through cracks and holes in the building and gaps and voids in ceilings, walls, and floors—and in the insulation.

Radiation heat transfer occurs between objects that are not touching. The most common example of this is the way the sun heats the earth. The sun warms the earth without warming the space between the sun and the earth. An example in our industry is the heat that you feel from a radiant heater or large cast iron radiator.

Types of Heat	
Sensible Heat	Sensible heat is any heat transfer that causes a change in temperature without causing a change of state. Sensible heat can be measured with a dry bulb thermometer. For example, when the temperature is increased over a heating coil, the temperature differential is sensible heat.
Latent Heat	Latent heat is the amount of heat required to cause a change of state. In a boiler system, this would be the amount of heat added to water to cause it to change from water to steam. It requires 970.4 Btus to raise 1 pound of water at 212 degrees f to 1pound of steam. The latent heat that is added to change water to steam is also given back when the steam condenses in the radiator or coil.
Total Heat	Total heat is the sum of the sensible and latent heat in an exchange process. It is sometimes called enthalpy.

Heat Transfer Coefficients			
Heating Medium	Transfer Material	Substance to be heated	$Btu/ft^2\ hr\ ^0F$
Water	Cast Iron	Air or Gas	1.4
Water	Mild Steel	Air or Gas	2.0
Water	Copper	Air or Gas	2.25
Water	Cast Iron	Water	40-50
Water	Mild Steel	Water	60-70
Water	Copper	Water	62-80
Air	Cast Iron	Air	1.0
Air	Mild Steel	Air	1.4
Steam	Cast Iron	Air	2.0
Steam	Mild Steel	Air	2.5
Steam	Copper	Air	3.0
Steam	Cast Iron	Water	160
Steam	Mild Steel	Water	185
Steam	Copper	Water	205
Steam	Stainless Steel	Water	120
The above are average coefficients for still fluids			

Did you know that the toilet is flushed more times during the halftime of the Super Bowl than at any other time of year?

Piping

Number of smaller pipes that equal One larger pipe

Pipe Size	1/2	3/4	1	1 1/4	1 1/2	2	2 1/2	4	6
1/2	1.00	2.27	4.88	10.00	15.80	31.70	52.90	205	620
3/4		1.00	2.05	4.30	6.97	14.00	23.30	90	273
1			1.00	2.25	3.45	6.82	11.40	44	133
1 1/4				1.00	1.50	3.10	5.25	19	68
1 1/2					1.00	2.00	3.34	13	39
2						1.00	1.67	6.50	19.60
2 1/2							1.00	3.87	11.70
4								1.00	3.02
6									1.00

Standard Nipples & Pipe Sizing Schedule 40					
Pipe Size	**Outside Diameter (O.D.)**	**Circumference**	**Pipe Size**	**Outside Diameter (O.D.)**	**Circumference**
1/8"	0.405"	1.272"	**2 1/2"**	2.875"	9.032"
¼ "	0.540"	1.696"	**3"**	3.500"	10.995"
3/8"	0.675"	2.121"	**4"**	4.500"	14.137"
½"	0.840'	2.639"	**5"**	5.563"	17.476"
¾"	1.050"	3.299"	**6"**	6.625"	20.812"
1"	1.315"	4.131"	**8"**	8.625"	27.095"
1 ¼"	1.660"	5.215"	**10"**	10.750"	33.771"
1 ½"	1.900"	5.969"	**12"**	12.750"	40.054"
2"	2.375"	7.461"			

Standard Copper Tubing Type K,L,M					
Pipe Size	Outside Diameter (O.D.)	Circumference	**Pipe Size**	Outside Diameter (O.D.)	Circumference
½"	0.625"	1.964"	**3"**	3.125"	9.817"
¾"	0.875"	2.749"	**4"**	4.125"	12.959"
1"	1.125"	3.534"	**6"**	6.127"	12.248"
1 ¼"	1.375"	4.319"	**8"**	8.125"	25.525"
1 ½"	1.625"	5.105"	**10"**	10.125"	31.808"
2"	2.125"	6.675"	**12"**	12.750"	40.054"
2 1/2"	2.625"	8.246"			

PIPING RESISTANCE

Typical Friction head for pipe is 4.2' per 100 feet @500 Milinches restriction per foot

Frictional Allowance for Fittings In Feet of Pipe					
Length to be added in feet					
Size of Fittings Inches	**90^0 Ell**	**Side Outlet Tee**	**Gate Valve**	**Globe Valve**	**Angle valve**
1/2"	1.3	3	0.3	14	7
3/4'	1.8	4	0.4	18	10
1"	2.2	5	0.5	23	12
1 1/4"	3.0	6	0.6	29	15
1 1/2"	3.5	7	0.8	34	18
2"	4.3	8	1.0	46	22
2 1/2"	5.0	11	1.1	54	27
4"	9	18	1.9	92	45
6"	13	27	2.8	136	67
8"	17	35	3.7	180	92
10"	21	45	4.6	230	112

Standard Nipples & Pipe Sizing Schedule 40		
Pipe Size	**Outside Diameter (O.D.)**	**Circumference**
1/8"	0.405"	1.272"
¼ "	0.540"	1.696"
3/8"	0.675"	2.121"
½"	0.840'	2.639"
¾"	1.050"	3.299"
1"	1.315"	4.131"
1 ¼"	1.660"	5.215"
1 ½"	1.900"	5.969"
2"	2.375"	7.461"
2 1/2"	2.875"	9.032"
3"	3.500"	10.995"
4"	4.500"	14.137"
5"	5.563"	17.476"
6"	6.625"	20.812"
8"	8.625"	27.095"
10"	10.750"	33.771"
12"	12.750"	40.054"

Standard Copper Tubing Type K,L,M		
Pipe Size	**Outside Diameter (O.D.)**	**Circumference**
½"	0.625"	1.964"
¾"	0.875"	2.749"
1"	1.125"	3.534"
1 ¼"	1.375"	4.319"
1 ½"	1.625"	5.105"
2"	2.125"	6.675"
2 1/2"	2.625"	8.246"
3"	3.125"	9.817"
4"	4.125"	12.959"
6"	6.127"	12.248"
8"	8.125"	25.525"
10"	10.125"	31.808"
12"	12.750"	40.054"

Frictional Allowance for Fittings In Feet of Pipe					
			Length to be added in feet		
Size of Fittings Inches	**90^0 Ell**	**Side Outlet Tee**	**Gate Valve**	**Globe Valve**	**Angle valve**
1/2"	1.3	3	0.3	14	7
3/4'	1.8	4	0.4	18	10
1"	2.2	5	0.5	23	12
1 1/4"	3.0	6	0.6	29	15
1 1/2"	3.5	7	0.8	34	18
2"	4.3	8	1.0	46	22
2 1/2"	5.0	11	1.1	54	27
4"	9	18	1.9	92	45
6"	13	27	2.8	136	67
8"	17	35	3.7	180	92
10"	21	45	4.6	230	112

PIPING EXPANSION

Calculating the linear expansion of pipe carrying steam or hot water. If you would like to calculate the expansion or lengthening of a pipe when it has steam or hot water inside, try this formula

$$E = Constant * Temperature\ rise$$

E = Expansion in inches per 100 feet of pipe C = Constant

Constant = Coefficient of expansion per 100 Ft pipe

Metal	Constant
Steel	0.00804
Wrought Iron	0.00816
Cast Iron	0.00780
Copper or Brass	0.01140

For Example

What is the expansion of 100 feet of copper tubing that will heat water from 50^{0}F to 180^{0}F?

$$E = Constant * Temperature\ rise$$

$$E = 0.01140 * (180 - 50) Temp\ rise$$

$$1.482" = 0.01140 * 130$$

The expansion is 1.482" for every 100 feet of copper tubing.

What is the expansion of 100 feet of steel pipe that will heat water from 50^{0}F to 180^{0}F?

$$E = Constant * Temperature\ rise$$

$$E = 0.00804 * (180 - 50) Temp\ rise$$

$$1.045" = 0.00804 * 130$$

The expansion is 1.045" for every 100 feet of steel pipe.

Copper expands at 39% greater rate than steel

Thermal Expansion of Piping Material in inches per 100 feet from 32 Deg F

Temperature Rise Deg F	Carbon & Carbon Moly Steel	Cast Iron	Copper
	Inches	Inches	Inches
32°F - 32°F	0	0	0
32°F - 100°F	0.5	0.5	0.8
32°F - 150°F	0.8	0.8	1.4
32°F - 200°F	1.2	1.2	2.0
32°F - 250°F	1.7	1.5	2.7
32°F - 300°F	2.0	1.9	3.3
32°F - 350°F	2.5	2.3	4.0
32°F - 400°F	2.9	2.7	4.7
32°F - 500°F	3.8	3.5	6.0
32°F - 600°F	4.8	4.4	7.4
32°F - 700°F	5.9	5.3	9.0

BTU Transportation Options

If you would like to deliver 1,000,000 Btu's to the space, you would need

Transportation Method	Size
Ductwork	**70" x 46" duct**
Hydronic piping	**3"**
Low Presssure Steam less than 40 fps	**6"**

Pipe Insulation

Heat Loss from Pipe

Heat Losses from Uninsulated Horizontal Steel Pipe *Btu per hour per linear foot at 70 DegreesF room temperature*		
Nom Pipe Size	Hot Water (180 Deg F)	Steam 5 PSIG
½	60	96
¾	73	118
1	90	144
1 ¼	112	170
1 ½	126	202
2	155	248
2 ½	185	296
3	221	355
4	279	448

Heat Losses from Uninsulated Horizontal Copper Pipe Btu per hour per linear foot at 70 Degrees F room temperature	
Nom Pipe Size	Hot Water (180 Deg F)
½	33
¾	45
1	55
1 ¼	66
1 ½	77
2	97
2 ½	117
3	136
4	174

Avg. Loss from Insulated steel pipe Btu/ Linear foot @ 70 Degrees F			
Pipe Size Inches	**Insulation Thickness**	**175⁰F**	**225⁰F**
½	1"	0.150	0.157
¾	1"	0.172	0.177
1	1"	0.195	0.200
	1 ½"	0.165	0.167
1 ¼"	1"	0.250	0.282
	1 ½"	0.170	0.193
1 ½"	1"	0.247	0.255
	1 ½"	0.205	0.210
2"	1"	0.290	0.297
	1 ½"	0.235	0.240
	2"	0.200	0.205
2 ½"	1"	0.330	0.340
	1 ½"	0.265	0.270
	2"	0.225	0.230
3"	1"	0.385	0.395
	1 ½"	0.305	0.312
	2"	0.257	0.263
4"	1"	0.470	0.480
	1 ½"	0.370	0.379
	2"	0.308	0.315

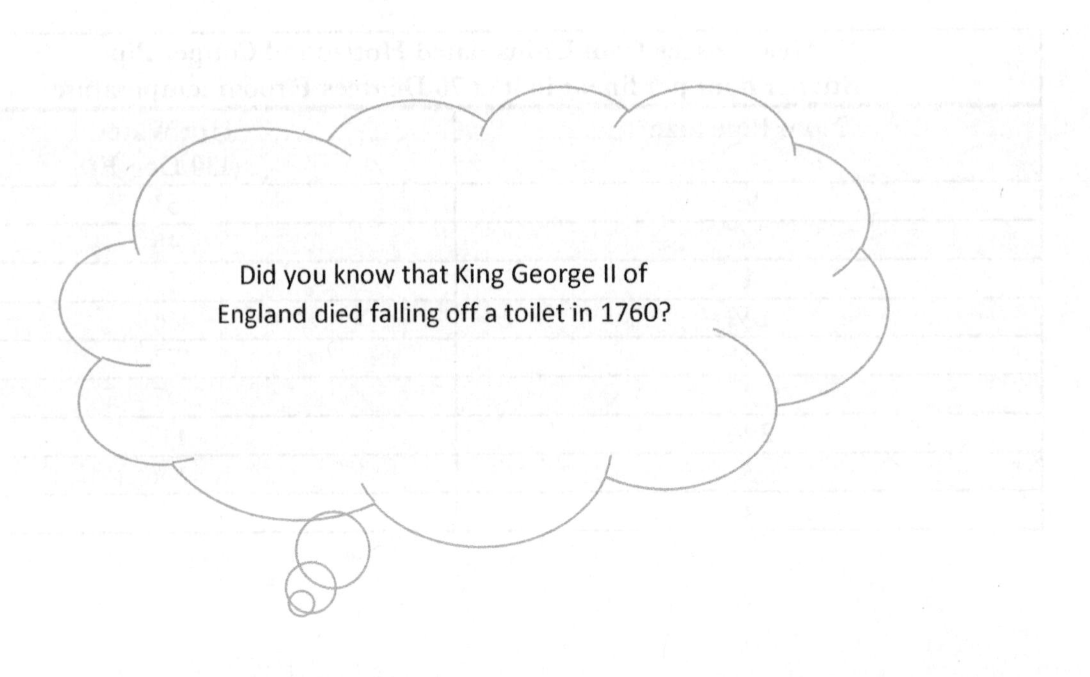

pH Scale

pH	Equivalent	If pH was measured with dollars
PH = 0	Battery Acid	-$100,000,000
PH = 1	Hydrochloric Acid	-$10,000,000
PH = 2	Lemon Juice, Vinegar	-$1,000,000
PH = 3	Grapefruit, Orange Juice	-$100,000
PH = 4	Acid Rain, Tomato Juice	-$10,000
PH = 5	Black Coffee	-$1,000
PH = 6	Urine, Saliva	-$100
PH = 7	"Pure" Water	$10
PH = 8	Sea Water	$100
PH = 9	Baking Soda	$1,000
PH = 10	Milk of Magnesia	$10,000
PH = 11	Ammonia Solution	$100,000
PH = 12	Soapy Water	$1,000,000
PH = 13	Bleaches, Oven Cleaner	$10,000,000
PH = 14	Liquid Drain Cleaner	$100,000,000

Acidic pH at 5-6, could cause corrosion. High alkaline between 8-10 causes scale deposits on hot transfer surfaces. The column on the far right shows the equivalent readings if pH was measured with dollars.

NOTE: Typical pH from condensing boiler flues are between 2.9 to 4.0.

Fuel Information

Comparative Fuel Values

To get 1,000,000 Btu's you need the following

Fuel Source	1,000,000 Btu's
Natural Gas @ 1000 Btu/ cu ft	1,000 Cu ft
Coal @ 12,000 Btu/ lb	83.333 Lb
Propane @ 91,600 Btu/ gal	10.917 Gal
Gasoline @ 125,000 Btus/gal	8.000 Gal
Fuel Oil #2 @ 140,000 Btus/gal	7.194 Gal
Fuel Oil #6 @ 150,000 Btus/gal	6.666 Gal
Electricity @ 3,412 Btu/kWh	293.083 Kwh

Average Btu Content of Common Fuels

Fuel Type	Number of Btu/ Unit
Fuel Oil #2	140,000 / Gallon
Fuel Oil #6	150,000 / Gallon
Butane	3,200 Btu's / CF
	21,500 Btus/ pound
	102,400 / Gallon
Natural Gas	1,025,000/ 1,000 cubic feet
Propane	91,330/ gallon
Coal	28,000,000 per ton
Electricity	3,412/ KWH
Wood Mixed	14,000,000/ cord or 3,500 / pound
Wood, Air Dried	20,000,000/ cord or 8,000 / pound
Kerosene	135,000 / gallon
Pellets	16,500,000/ton

Misc. Fuel Information

Natural Gas	
1 Cu ft Natural Gas =	1,000 Btus
1 MCF =	1,000,000 Btus
	1 MMBTU
	1 MCF
	1,000 Cu Ft.
	10 CCF
	10 Therms
1 Dekatherm	1 MCF
	10 Therms
	1,000,000 Btus
1 Therm =	100,000 Btus or 100 MBTU
	0.1 MCF
	1 CCF
	100 Cu. Feet
1 CCF =	1,000 Cu Ft
	100 Therm
Propane	
1 gallon =	92,000 Btus
1 Cu Foot =	2,250 Btus
#2 Fuel Oil	
1 Gallon =	140,000 Btus

Gas Pipe Line Sizing

	Pipe Length			
Steel Pipe	**10 Feet**	**20 Feet**	**40 feet**	**80 Feet**
Size	Capacity in Cubic Feet per hour			
½"	120	85	60	42
¾"	272	192	136	96
1"	547	387	273	193
11/4"	1,200	849	600	424
1 ½"	1,860	1,316	930	658
2"	3,759	2,658	1,880	1,330
2 ½"	6,169	4,362	3,084	2,189
4"	23,479	16,602	11,740	8,301

Each cubic foot of gas roughly equals 1,000 Btuh

Boiler Gas Consumption

Btuh	Cu Feet/Hr	Cubic Feet/ Min	Cubic Feet/ Sec
20,000,000	20,000	333.33	5.56
19,000,000	19,000	316.67	5.28
18,000,000	18,000	300.00	5.00
17,000,000	17,000	283.33	4.72
16,000,000	16,000	266.67	4.44
15,000,000	15,000	250.00	4.17
14,000,000	14,000	233.33	3.89
13,000,000	13,000	216.67	3.61
12,000,000	12,000	200.00	3.33
11,000,000	11,000	183.33	3.06
10,000,000	10,000	166.67	2.78
9,000,000	9,000	150.00	2.50
8,000,000	8,000	133.33	2.22
7,000,000	7,000	116.67	1.94
6,000,000	6,000	100.00	1.67
5,000,000	5,000	83.33	1.39
4,000,000	4,000	66.67	1.11
3,000,000	3,000	50.00	0.83
2,000,000	2,000	33.33	0.56
1,000,000	1,000	16.67	0.28
Based upon 1,000 Btu's per cubic foot			

Gas Pipe Threads The following table shows the length and approximate number of threads on gas lines.

Iron Pipe Size (Inches)	Approximate Length of Threaded Portion	Approximate number of threads to be cut
1/2	¾	10
¾	¾	10
1	7/8	10
1 ¼	1	11
1 ½	1	11
2	1	11
2 1/2	1 ½	12
3	1 ½	12
4	1 5/8	13

Corrugated Stainless Steel Tubing CSST Sizing
Rules of Thumb for EHD Sizing EHD = Equivalent Hydraulic Diameter

EHD	Pipe Size	EHD	Pipe Size
15	3/8"	37	1 1/4"
19	1/2"	46	1 1/2"
25	3/4"	62	2"
31	1"		
Please check with the manufacturer to verify their sizing			

Inlet Pressure		Pressure Drop		Specific Gravity	
Less than 2 psi		0.6"		0.60	
Tube Size	Length				
EHD	**5**	**10**	**15**	**20**	**25**
	Capacity in Cu ft per Hour				
13	46	32	25	22	19
15	63	44	35	31	27
18	115	82	66	58	52
19	134	95	77	67	60
23	225	161	132	116	104
25	270	192	157	137	122
30	471	330	267	231	206
31	546	383	310	269	240
37	895	639	524	456	409
46	1792	1260	1030	888	793
48	2070	1470	1200	1050	936
60	3660	2930	2400	2080	1860
EHD	**30**	**40**	**50**	**60**	**70**
13	18	15	13	12	11
15	25	21	19	17	16
18	47	41	37	34	31
19	55	47	42	38	36
23	96	83	75	68	63
25	112	97	87	80	74
30	188	162	144	131	121
31	218	188	168	153	141
37	374	325	292	267	248
46	723	625	559	509	471
48	856	742	665	608	563
60	1520	1320	1180	1080	1000

Gas Pressure Comparison

Inches Hg	Ounces	PSI
0.1	0.05	0.003
0.2	0.12	0.007
0.4	0.23	0.01
0.6	0.35	0.02
0.8	0.46	0.028
1	0.58	0.036
2	1.15	0.072
3	1.73	0.108
4	2.31	0.144
5	2.89	0.18
6	3.46	0.216
7	4.04	0.252
8	4.62	0.288
9	5.20	0.324
10	5.78	0.36
11	6.35	0.396
12	6.93	0.432
13	7.51	0.468
14	8.09	0.504
15	8.67	0.54
16	9.24	0.576
17	9.82	0.612
18	10.4	0.648
19	10.98	0.684
20	11.56	0.72
21	12.13	0.756
22	12.71	0.792
23	13.29	0.828
24	13.87	0.864
25	14.45	0.9
26	15.02	0.936
27	15.60	0.972
28	16.18	1.008
29	16.76	1.044
30	17.34	1.08
31	17.91	1.16
32	18.49	1.152
33	19.07	1.188

Pressure Conversion Chart

Inches H^2O to PSI
28" W.C. = 1 psi

Inches H^2O	PSI		Inches H^2O	PSI
0.1	0.0036		**15**	0.5414
0.2	0.0072		**16**	0.5774
0.4	0.0144		**17**	0.6136
0.6	0.0216		**18**	0.6496
0.8	0.0289		**19**	0.6857
1	0.0361		**20**	0.7218
2	0.0722		**21**	0.7579
3	0.1083		**22**	0.7940
4	0.1444		**23**	0.8301
5	0.1804		**24**	0.8662
6	0.2165		**25**	0.9023
7	0.2526		**26**	0.9384
8	0.2887		**27**	0.9745
9	0.3248		**28**	1.010
10	0.3609		**29**	1.047
11	0.3970		**30**	1.083
12	0.4331		**31**	1.191
13	0.4692		**32**	1.155
14	0.5053		**33**	1.191

Convert Pascals to Inches of Water - 0.004014631332 x Pascals = Inches W. C.
Comvert Inches W.C. to Pascals – 249.088875 x Inches W.C. = Pascals

Pascal	Inches H^2O	**Pascal**	Inches H^2O	**Pascal**	Inches H^2O	Pascal	Inches H^2O
1	0.00	**65**	0.26	**130**	0.52	**195**	0.78
5	0.02	**70**	0.28	**135**	0.54	**200**	0.80
10	0.04	**75**	0.30	**140**	0.56	**210**	0.82
15	0.06	**80**	0.32	**145**	0.58	**215**	0.84
20	0.08	**85**	0.34	**150**	0.60	**220**	0.86
25	0.10	**90**	0.36	**155**	0.62	**225**	0.88
30	0.12	**95**	0.38	**160**	0.64	**230**	0.90
35	0.14	**100**	0.40	**165**	0.66	**235**	0.92
40	0.16	**105**	0.42	**170**	0.68	**240**	0.94
45	0.18	**110**	0.44	**175**	0.70	**245**	0.96
50	0.20	**115**	0.46	**180**	0.72	**250**	0.98
55	0.22	**120**	0.48	**185**	0.74	**255**	1.00
60	0.24	**125**	0.50	**190**	0.76	**260**	1.02

Orifice Capacities for Natural Gas

1,000 Btu per cubic foot
Manifold pressure 3 ½" Water Column

Wire Gauge Drill Size	Decimal Inches	Rate Cu Ft / Hr	Rate Btu/Hr
70	0.028	1.34	1,340
68	0.0310	1.65	1,650
66	0.0330	1.80	1,870
64	0.0360	2.22	2,250
62	0.0380	2.45	2,540
60	0.0400	2.75	2,750
58	0.0420	3.50	3,050
56	0.0465	3.69	3,695
54	0.055	5.13	5,125
52	0.0635	6.92	6,925
50	0.0700	8.35	8,350
48	0.0760	9.87	9,875
46	0.0810	11.25	11,250
44	0.0860	12.62	12,625
42	0.0935	15.00	15,000
40	0.0980	16.55	16,550
38	0.1015	17.70	17,700
36	0.1065	19.50	19,500
34	0.1110	21.05	12,050
32	0.1160	23.70	23,075
30	0.1285	28.50	28,500
28	0.1405	34.12	34,125
26	0.1470	37.25	37,250
24	0.1520	38.75	38,750
22	0.1570	42.50	42,500
20	0.1610	44.75	44,750

Flue Information

Flue Information

Vent Categories				
	I	II	III	IV
Vent Pressure	Negative	Negative	Positive	Positive
Temperature	$>275^0$	$<275^0$	$>275^0$	$<275^0$
Efficiency	$<84\%$	$>84\%$	$<84\%$	$>84\%$
Gas tight	No	No	Yes	Yes

The Products of Combustion Produced When One Cubic Foot of Gas is Burned	
One Cubic Foot of Gas Burned Produces	8 Cubic feet of nitrogen
	2 Cubic feet of water vapor
	1 Cubic Foot of Nitrogen

Typical Vent Temperature Ranges		
Venting Material	Temperature Ratings	Fuel
AL29-4C Stainless	0 - 480^0 F	Gas
B and BW Vent	0 - 550^0 F	Gas
L Vent	0 - 1,000^0 F	Oil
Factory Built Chimney	500^0 - 2,200^0 F	Oil/Gas
Masonry Chimney	360^0 - 1,800^0 F	Oil/Gas
Verify with manufacturer		

Condensing & Ignition Temperature of Various Fuels

Fuel	Condensing Temperature	Ignition Temperature
Natural Gas	250 ^{0}F	1,163 ^{0}F
#2 Oil	275 ^{0}F	600 ^{0}F
#6 Oil	300 ^{0}F	765 ^{0}F
Coal	325 ^{0}F	850 ^{0}F
Wood	400 ^{0}F	540-1,100 ^{0}F

Minimum Flue Gas Temperatures for Category 1 Boilers

Fuel	Minimum Flue Temperature
Natural Gas	265°F plus 1/2°F for each foot of stack or breeching, including horizontal and vertical runs
#2 Fuel Oil	240°F plus 1/2°F for each foot of stack or breeching, including horizontal and vertical runs

Acid Rain and Stack Temperature

Fuel	Acid Dew Point Temperature	Minimum Allowable Stack Temperature
Natural Gas	150	250
#2 Fuel Oil	180	275

Typical Boiler Exhaust Velocity	
Equipment Type	Typical Exhaust Velocity ft/s
Boiler with On Off Burner	16-26
Boiler with two step burner	31-49
Boiler with modulating burner	49-82
Minimum to keep surface free from soot	9.8-13

ELECTRICAL

Amp of Copper Wire Types
Single wire in open air

Wire Size AWG	TH UF	FEPW, RH, RHW, TWH, THWN, ZW, THHW, XHHW	USE-2, XHH, XHHW, TBS, SA, SIS, FEP, MI, RHW-2, THHN, ZW-2, THWN-2, FEPB, RHH, THHW, THW-2
0000	300	360	405
000	260	310	350
00	225	265	300
0	195	230	260
1	165	195	220
2	140	170	190
3	120	145	165
4	105	125	140
6	80	95	105
8	60	70	80
10	40	50	55
12	30	35	40
14	25	30	35
16	-	-	24
18	-	-	18

Up to 86-degree ambient temperature

Amp of Copper Wire Types
Three wires in cable

Wire Size AWG	TH UF	FEPW, RH, RHW, TWH, THWN, ZW, THHW, XHHW	USE-2, XHH, XHHW, TBS, SA, SIS, FEP, MI, RHW-2, THHN, ZW-2, THWN-2, FEPB, RHH, THHW, THW-2
0000	195	230	260
000	165	200	225
00	145	175	195
0	125	150	170
1	110	130	150
2	95	115	130
3	85	100	110
4	70	85	95
6	55	65	75
8	40	50	55
10	30	35	40
12	25	25	30
14	20	20	25
16	-	-	18
18	-	-	14

Up to 86-degree ambient temperature

How to calculate Electrical Phase Imbalance

% of Voltage Imbalance =Maximum deviation from average ÷ average x 100

Take average of all 3 readings between all legs to get average. Most motors limit imbalance to 2%

% Imbalance	% Motor Winding Temperature Increase	% Imbalance	% Motor Winding Temperature Increase
2%	8%	**5%**	50%
3%	18%	**6%**	72%
4%	32%	**7%**	98%

Standard 24 Volt Thermostat Connections		
Terminal	Usage	Normal Colors
R or V	24 VAC power	Red
Rh or 4	24 VAC Heating Power	Red
Rc	24 VAC Cooling Power	Red
C	24 VAC Common	Black
Y	1st Stage Cooling	Yellow
Y2	2nd Stage Cooling	Blue or Orange
W	1st Stage Heat	White
W2	2nd Stage Heat	No Standard Color
G	Fan	Green

Ohm's Law

Volts =	$\sqrt{Watts \times Ohms}$	**Amperes =**	$\frac{Volts}{Ohms}$
	$\frac{Watts}{Amperes}$		$\frac{Watts}{Volts}$
	Amperes x Ohms		$\sqrt{\frac{Watts}{Ohms}}$
Ohms =	$\frac{Volts}{Amperes}$	**Watts =**	Volts x Amperes
	$Watts/Amperes^2$		$Amperes^2$ x Ohms
	$\frac{Volts^2}{Watts}$		$\frac{Volts^2}{Ohms}$

Electrical Equivalents Formulas	
Watt=	44.236 foot-pounds minute
	2,654.16 foot-pounds hour
	0.00134 hp
	3.414 Btu
	0.0035 lb of water evaporated per hour
	44.236 foot-pounds minute
	2,654.16 foot-pounds hour
Kilowatt=	44,235 foot-pounds minute
	1.34 H.P.
	0.955 BTU per second
	57.3 Btu per minute
	3,438 Btu per hour
	1,000 W
	1.34 hp
	3.53 lbs water evaporated per hr from and at 212 degrees F
	0.955 Btu's
	57.3 BTU per minute
	3,413 Btuh per hour
1 H.P.=	33,000 foot-pounds minute
	746 watts
	42.746 Btu per minute
	2,564.76Btu per hour
1 Btu=	772 ft lbs
	17.452 watts per minute
	0.2909 watts hour

Calculate Motor HP from Meter Readings	
DC Motors	$HP = \frac{V * A * Ef}{746}$
Single Phase AC Motors	$HP = \frac{V * A * Ef * PF}{746}$
Three Phase AC Motors	$HP = \frac{V * A * Ef * PF * 1.73}{746}$
V= Volts	
Ef= Motor Efficiency	
A= Amps	
PF= Power factor	

Full Load Amperes of Single Phase Motors			
HP	**RPM**	**115V**	**230V**
1/8	1725	2.8	1.4
	1140	3.4	1.7
	860	4.0	2.0
1/4	1725	4.6	2.3
	1140	6.15	3.07
	860	7.5	3.75
1/3	1725	5.2	2.6
	1140	6.25	3.13
	860	7.35	3.67
1/2	1725	7.4	3.7
	1140	9.15	4.57
	860	12.8	6.4
3/4	1725	10.2	5.1
	1140	12.5	6.25
	860	15.1	7.55
1	1725	13.0	6.5
	1140	15.1	7.55
	860	15.9	7.95

Full Load Amperes of Three Phase Motors			
HP	RPM	115V	230V
1/4	1725	0.95	0.48
	1140	1.4	0.7
	860	1.6	0.8
1/3	1725	1.19	0.6
	1140	1.59	0.8
	860	1.8	0.9
1/2	1725	1.72	0.86
	1140	2.15	1.08
	860	2.38	1.19
3/4	1725	2.46	1.23
	1140	2.92	1.46
	860	3.26	1.63
1	1725	3.19	1.6
	1140	3.7	1.85
	860	4.12	2.06
1 1/2	1725	4.61	2.31
	1140	5.18	2.59
	860	5.75	2.88
2	1725	5.98	2.99
	1140	6.5	3.25
	860	7.28	3.64
3	1725	8.70	4.35
	1140	9.25	4.62
	860	10.3	5.15
5	1725	14.0	7.0
	1140	14.6	7.3
	860	16.2	8.1
7 1/2	1725	20.3	10.2
	1140	20.9	10.5
	860	23.0	11.5

Typical Flame Safeguard Signals Siemens	
Gas Burner Controls	
Model	**Flame Signal**
LFL with UV sensor QRA Minimum 70 uA DC	100-450 µA DC Typical
LFL with Flame Rod Minimum 6 uA DC	20-100 uA DC
Oil Burner Controls	
Model	**Flame Signal**
LAL1 with photoresistive detector, QRB1	95-160 µA DC
LAL1 with blue-flame detector, QRC1	80-130µA DC
LAL2/LAL3 with photoresistive detector, QRB1	8-35 µA DC
LAL2/LAl3 with selenium photocell detector, RAR	6.5-30 µA DC
LAL4 with photoresistive detector, QRB1	95-160 µA DC
LAL4 with blue-flame detector, QRC1	80-130 µA DC

Flame Safeguard Definitions	
µA = Micro Amps	VDC = Volts DC
1 Micro Amp or µA = 0.000001 Amps	
1 Amp = 1,000,000 Micro Amps	
1 Amp = 1,000 Milliamps or mA	

Typical Flame Safeguard Signals Fireye & Honeywell	
Fireye	
Model	Average Flame Signal
UVM	4.0-5.5 VDC
TFM	14-17 VDC
D-10/20/30	16-25 VDC
E-100/ 110	20-80 VDC
E-100/E110 with EPD Programmer	4-10 VDC
M Series II	4-6 VDC
Micro M Series	4-10 VDC
Micro M Series w Display	20-80 VDC
Honeywell	
Model	Average Flame Signal
RA890	2-6 μA DC
R4795	2-6 μA DC
R7795	2-6 μA DC
R4140	2-6 μA DC
R4150	2-6 μA DC
BC7000	2-6 μA DC
RM7890	1.25-5 VDC
RM7895	1.25 VDC
RM7840	1.25 VDC
RM7800	1.25 VDC
S8600	1-5 μA DC

Flame Safeguard Definitions	
μA = Micro Amps	VDC = Volts DC
1 Micro Amp or μA = 0.000001 Amps	
1 Amp = 1,000,000 Micro Amps	
1 Amp = 1,000 Milliamps or mA	

MISC

Pressure Unit Conversions

Known	Desired Pressure Unit			
	Pounds Per sq In.	**Ounces Per Sq In.**	**Inches of Water**	**Feet of Water**
Centimeters of Water	0.0981	0.227	0.384	0.0328
Feet of water	0.433	6.94	12.0	0.883
Inches Mercury	0.491	7.86	13.6	1.13
Inches Water	0.0361	0.578	-------	0.0833
Ounces per Square Inch	0.0625	---------	1.73	0.144
Pounds per Sq Inch	----------	16.0	27.7	2.31

Common Fraction to Decimal to Millimeters		
Fraction	**Decimal**	**Millimeters**
1/16	0.0625	1.587
1/8	0.125	3.175
3/16	0.1875	4.762
1/4	0.250	6.350
5/16	0.3125	7.937
3/8	0.375	9.525
7/16	0.4375	11.113
1/2	0.50	12.700
9/16	0.5625	0.5625
5/8	0.625	15.875
11/16	0.6875	17.462
3/4	0.750	19.050
13/16	0.8125	20.637
7/8	0.875	22.225
15/16	0.9375	23.812
1	1.00	25.400

Conversion Tables

CONVERSION FACTORS	
Length & Area	
1 Mile =	1,760 yards
	5,280 feet
	63,360 inches
	1.609 Km
1 Foot =	0.3048 M
	30.48 Cm
	304.8 mm
1 Inch =	25,400 microns
1 acre =	43,560 Sq Ft
	4,840 Sq Yds
	0.4047 Hectares
1 Sq Mile =	640 Acres
1 Sq Yd =	9 Sq Ft
	1,296 Sq Inches
1 Sq Foot =	144 Square Inches
1 Cu Yard =	27 Cu Ft
	46,656 cu inches
	1,616 pints
	807.9 quarts
	764.6 Liters
1 Cu foot =	1,728 cubic inches
1 Gallon H2O =	8.333 Lbs
Diameter of Circle =	Circumference x 0.3188
Circumference of Circle =	Diameter x 3.1416
1 League =	3.0 Miles
Pressure	
1 lb steam =	1 lb H2O
14.7 psi =	33.95 ft H2O
	29.92 in Hg
	407.2 In wg
	2,116.8 Lbs/ Sq Ft
1 Psia =	Psig = 14.7
1 psi =	2.307 Ft H2O
	2.036 In Hg
	16 ounces
	27.7 In w.c.
1 ounce =	1.73 inches w.c
1 Ft H20 =	0.4335 psi
	62.43 lbs/ sq feet

Weight	
# =	Pounds of Pressure
1 Lb =	16 oz
	7,000 grains
	0.4536 Kg
1 ton =	2,000 lbs
	907 Kg
Liquid	
1 gallon =	4 quarts
	8 pints
	3.785 liters
	0.13368 Cu Feet
	231 Cu Inches
1 Liter =	0.2642 Gallons
	1.057 quarts
	2,113 pints
Speed	
1 MPH =	5280 ft / hr
	88 ft/min
	1.467 ft/sec
	0.868 Knots per Hr
1 Knot =	1.1515 MPH
Speed of Sound in Air =	1,128.5 ft/sec
	769.4 mph
Misc	
1 Barrel Oil =	42 gallons
Hours in a Year	8,760

Diameter to Circumference in Inches

Diameter	Circumference	Diameter	Circumference
12	37.70	**28**	87.96
14	43.98	**30**	94.25
16	50.27	**32**	100.53
18	56.55	**34**	106.81
20	62.83	**36**	113.10
22	69.12	**38**	119.38
24	75.40	**40**	125.66
26	81.68		

Pressure Unit Conversions

Known	Desired Pressure Unit			
	Pounds Per sq In.	**Ounces Per Sq In.**	**Inches of Water**	**Feet of Water**
Centimeters of Water	0.0981	0.227	0.384	0.0328
Feet of water	0.433	6.94	12.0	0.883
Inches Mercury	0.491	7.86	13.6	1.13
Inches Water	0.0361	0.578	-------	0.0833
Ounces per Square Inch	0.0625	---------	1.73	0.144
Pounds per Sq Inch	----------	16.0	27.7	2.31

Estimate Storage Tank Capacity in Gallons
Rectangular Tank
Sizing a Storage Tank
If using this formula for sizing a boiler feed or condensate tank, the capacity will be about 50% of the tank size.

Rectangular Storage Tanks

Length" x Width" x Height" divided by 231 = Gallons
Measure tank length, width and height in inches
Multiply Length x Width x Height
Divide that total by 231 = Gallons
The following are some tank sizes that are already calculated.

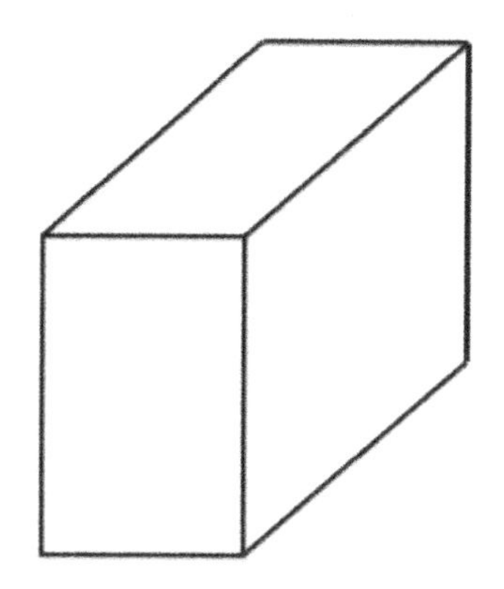

Rectangular Tank Height 12"				
	US Gallons			
	Width			
Length	12	18	24	36
12	7	11	15	22
18	11	17	22	34
24	15	22	30	45
30	19	28	37	56
36	22	34	45	67
42	26	39	52	79
48	30	45	60	90
60	37	56	75	112

Rectangular Tank Height 18"				
	US Gallons			
	Width			
Length	12	18	24	36
12	11	17	22	34
18	17	25	34	50
24	22	34	45	67
30	28	42	56	84
36	34	50	67	101
42	39	59	79	118
48	45	67	90	135
60	56	84	112	168

Rectangular Tank Height 24"				
	US Gallons			
	Width			
Length	12	18	24	36
12	15	22	30	45
18	22	34	45	67
24	30	45	60	90
30	37	56	75	112
36	45	67	90	135
42	52	79	105	157
48	60	90	120	180
60	75	112	150	224

Rectangular Tank Height 36"				
	US Gallons			
	Width			
Length	12	18	24	36
12	22	34	45	67
18	34	50	67	101
24	45	67	90	135
30	56	84	112	168
36	67	101	135	202
42	79	118	157	236
48	90	135	180	269
60	112	168	224	337

Rectangular Tank Height 48"				
	US Gallons			
	Width			
Length	12	18	24	36
12	30	45	60	90
18	45	67	90	135
24	60	90	150	269
30	75	112	150	224
36	90	135	180	269
42	105	157	209	314
48	120	180	239	359
60	150	224	299	449

Estimate Storage Tank Capacity in Gallons
Round Tanks

Multiply ½ Tank Diameter by itself
Multiply that by 3.146 x Length of Tank in inches
Divide by 231 = Gallons of water
For example, you have a 24" diameter tank that is 36" long
½ of 24" = 12" x 12" = 144
144 x 3.146 x 36 = 16,308
16,308 divided by 231 = 70.59 Gallons

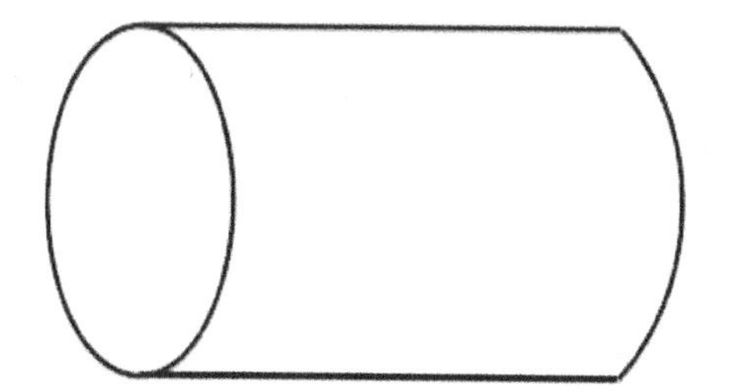

Length (feet)	Circular Tank Inside Diameter (Inches) US Gallons				
	18	**24**	**30**	**36**	**42**
1	1.1	1.96	3.06	4.41	5.99
2	26	47	73	105	144
2.5	33	59	91	131	180
3	40	71	100	158	216
3.5	46	83	129	184	252
4	53	95	147	210	288
4.5	59	107	165	238	324
5	66	119	181	264	360
5.5	73	130	201	290	396
6	79	141	219	315	432
6.5	88	155	236	340	468
7	92	165	255	368	504
7.5	99	179	278	396	540
8	106	190	291	423	576
9	119	212	330	476	648
10	132	236	366	529	720
12	157	282	440	634	864
14	185	329	514	740	1008

Circular Tank Inside Diameter (Inches)					
US Gallons					
Length (feet)	**48**	**54**	**60**	**66**	**72**
1	7.83	9.91	12.24	14.41	17.62
2	188	238	294	356	423
2.5	235	298	367	445	530
3	282	357	440	534	635
3.5	329	416	513	623	740
4	376	475	586	712	846
4.5	423	534	660	800	852
5	470	596	734	899	1057
5.5	517	655	808	978	1163
6	564	714	880	1066	1268
6.5	611	770	954	1156	1374
7	658	832	1028	1244	1480
7.5	705	889	1101	1355	1586
8	752	949	1175	1424	1691
9	846	1071	1322	1599	1903
10	940	1189	1463	1780	2114
12	1128	1428	1762	2133	2537
14	1316	1666	2056	2490	2960

Metric Liquid	
Metric	**U.S.**
3.7854 L	1 Gallon
0.946 L	1 Quart
0.473 L	1 Pint
1 L	0.264 Gallons
1 L	33.814 Ounces
29.576 ml	1 Fluid Ounce
236.584 ml	1 Cup
Metric Length	
Metric	U.S.
1m	39.37 inches
1 m	3.28 feet
1 m	1.094 yards
1 m	0.0016 miles
1.609 km	1 mile
25.4 mm	1 inch
2.54 cm	1 inch
304.8 mm	1 foot
1 mm	0.03937 inches
1 cm	0.3937 inches
1 dm	3.937 inches
Metric Pressure	
6.8947 kPa	1 pound per sq in (psi)
9.794 kPa	1m column of water
1 kPa	10.2 cm of water
1.3332 kPa	1 cm column of water
3.3864 kPa	1 inch of mercury (in Hg)
8 kPa	6 cm of mercury
Metric Conversions	
KJ/Hr =	Btu/h x 1.055
CMM =	CFM x 0.02832
LPM =	GPM x 3.785
Kj/Lb.=	Btu/Lb x 2.326
Meters =	Feet x 0.3048
Sq Meters =	Sq Feet x 0.0929
Cu. Meters =	Cu. Feet x 0.02832
Kg =	Pounds x 0.4536
Kg/Cu. Meter =	Pounds. Cu Feet x 16.017
Cu. Meters/ Kg =	Cu. Ft/ Pound x 0.0624

Fahrenheit to Celsius					
F	C	**F**	C	**F**	C
1	-17.2	**41**	5.0	**81**	27.2
2	-16.7	**42**	5.6	**82**	27.8
3	-16.1	**43**	6.1	**83**	28.3
4	-15.6	**44**	6.7	**84**	28.9
5	-15.0	**45**	7.2	**85**	29.4
6	-14.4	**46**	7.8	**86**	30.0
7	-13.9	**47**	8.3	**87**	30.6
8	-13.3	**48**	8.9	**88**	31.1
9	-12.8	**49**	9.4	**89**	31.7
10	-12.1	**50**	10.0	**90**	32.2
11	-11.7	**51**	10.6	**91**	32.8
12	-11.1	**52**	11.1	**92**	33.3
13	-10.6	**53**	11.7	**93**	33.9
14	-10.0	**54**	12.2	**94**	34.4
15	-9.4	**55**	12.8	**95**	35.0
16	-8.9	**56**	13.3	**96**	35.6
17	-8.3	**57**	13.9	**97**	36.1
18	-7.8	**58**	14.4	**98**	36.7
19	-7.2	**59**	15.0	**99**	37.2
20	-6.7	**60**	15.6	**100**	37.8
21	-6.1	**61**	16.1	**101**	38.3
22	-5.6	**62**	16.7	**102**	38.8
23	-5.0	**63**	17.2	**103**	39.4
24	-4.4	**64**	17.8	**104**	40
25	-3.9	**65**	18.3	**105**	40.6
26	-3.3	**66**	18.9	**106**	41.1
27	-2.8	**67**	19.4	**107**	41.6
28	-2.2	**68**	20.0	**108**	42.2
29	-1.7	**69**	20.6	**109**	42.8
30	-1.1	**70**	21.1	**110**	43.3
31	-0.6	**71**	21.7	**111**	43.8
32	0	**72**	22.2	**112**	44.4
33	0.6	**73**	22.8	**113**	45
34	1.1	**74**	23.3	**114**	45.6
35	1.7	**75**	23.9	**115**	46.1
36	2.2	**76**	24.4	**116**	46.6
37	2.8	**77**	25.0	**117**	47.2
38	3.3	**78**	25.6	**118**	47.7
39	3.9	**79**	26.1	**119**	48.3
40	4.4	**80**	26.7	**120**	48.9

Fahrenheit to Celsius					
F	C	**F**	C	**F**	C
121	49.4	**161**	71.7	**201**	93.9
122	50	**162**	72.2	**202**	94.4
123	50.5	**163**	72.8	**203**	95
124	51.1	**164**	73.3	**204**	95.6
125	51.7	**165**	73.9	**205**	96.1
126	52.2	**166**	74.4	**206**	96.7
127	52.8	**167**	75	**207**	97.2
128	53.3	**168**	75.6	**208**	97.8
129	53.9	**169**	76.1	**209**	98.3
130	54.4	**170**	76.7	**210**	98.9
131	55	**171**	77.2	**211**	99.4
132	55.6	**172**	77.8	**212**	100
133	56.1	**173**	78.3	**213**	100.6
134	56.7	**174**	78.9	**214**	101.1
135	57.2	**175**	79.4	**215**	101.7
136	57.8	**176**	80	**216**	102.2
137	58.3	**177**	80.6	**217**	102.8
138	58.9	**178**	81.1	**218**	103.3
139	59.4	**179**	81.7	**219**	103.9
140	60	**180**	72.2	**220**	104.4
141	60.6	**181**	82.8	**221**	105
142	61.1	**182**	83.3	**222**	105.5
143	61.7	**183**	83.9	**223**	106.1
144	62.2	**184**	84.4	**224**	106.7
145	62.8	**185**	85	**225**	107.2
146	63.3	**186**	85.6	**226**	107.8
147	63.9	**187**	86.1	**227**	108.3
148	64.4	**188**	86.7	**228**	108.9
149	65	**189**	87.2	**229**	109.4
150	65.6	**190**	87.8	**230**	110
151	66.1	**191**	88.3	**231**	110.5
152	66.7	**192**	88.9	**232**	111.1
153	67.2	**193**	89.4	**233**	111.6
154	67.8	**194**	90	**234**	112.2
155	68.3	**195**	90.6	**235**	112.8
156	68.9	**196**	91.1	**236**	113.3
157	69.4	**197**	91.7	**237**	113.9
158	70	**198**	92.2	**238**	114.4
159	70.56	**199**	92.8	**239**	115
160	71.1	**200**	93.3	**240**	115.5

Hydronic Heating Formulas & Rules of Thumb

Velocity Calculation

Typical Hydronic Velocity = 2 -4.5 Feet per second in occupied areas. Slightly higher in unoccupied areas. Flows above 6 feet per second could erode copper. Flows below 2 could allow air to be trapped in the piping and cause air locked systems.

To find Fluid Velocity

$$Feet\ per\ Second\ \frac{0.408\ x\ GPM}{(Pipe\ Diameter\ Inches)^2}$$

$$Gallons\ per\ Minute\ GPM = (Pipe\ Diameter\ Inches)^2 \div FPS$$

Feet per second FPS to Miles per Hour MPH			
Feet per second	**MPH**	**Feet per second**	**MPH**
1	0.68	**9**	6.13
2	1.37	**10**	6.82
3	2.05	**11**	7.50
4	2.73	**12**	8.18
5	3.40	**13**	8.87
6	4.09	**14**	9.55
7	4.77	**15**	10.23
8	5.45	**16**	10.91

This table will allow you to calculate the capacity in gallons per minute at different velocities.

For example, if you have a 2" pipe and want to keep the velocity at 4 feet per second, the pipe will be capable of running 54 gallons per minute.

Pipe GPM @ Various Velocities Feet per Second					
	FPM = Feet per Minute				
Pipe Size Inches	2 FPM	3 FPM	4 FPM	5 FPM	6 FPM
	Gallons per Minute				
1/2"	2	4	5	12	22
3/4"	4	6	8	21	38
1"	7	10	14	35	62
1 1/4"	12	18	24	60	108
1 1/2"	16	18	33	81	147
2"	27	24	54	135	242
3"	59	89	118	296	533
4"	102	153	204	510	918
6"	231	347	463	1,157	2,082
8"	400	600	800	2,000	3,599
10"	631	946	1,261	3,153	5,675
12"	895	1,343	1,791	4,476	8,058
14"	1,083	1,624	2,165	5,413	9,744
16"	1,413	2,120	2,825	7,065	12,717
18"	1,789	2,684	3,579	8,947	16,104
20"	2,222	3,333	4,444	11,110	19,998
24"	3,216	4,824	6,432	16,080	28,945

This table will help you calculate the velocity of different flows in different pipe sizes.

For example, if we have 100 gallons per minute flowing through a 4" pipe, the velocity will be 2.55 feet per second. The shaded areas are outside the normal velocities of 2-6 feet per second in hydronic pipes.

Convert Flow (GPM) to Velocity (FPS)

GPM	Pipe Diameter in Inches					
	2	**4**	**6**	**8**	**10**	**12**
	FEET PER SECOND					
5	0.51	0.13	0.06	0.03	0.02	0.01
10	1.02	0.26	0.11	0.06	0.04	0.03
15	1.53	0.38	0.17	0.10	0.06	0.04
20	2.04	0.51	0.23	0.13	0.08	0.06
30	3.06	0.77	0.34	0.19	0.12	0.09
40	4.08	1.02	0.45	0.26	0.16	0.11
50	5.10	1.28	0.57	0.32	0.20	0.14
60	6.12	1.53	0.68	0.38	0.24	0.17
70	7.14	1.79	0.79	0.45	0.29	0.20
80	8.16	2.04	0.91	0.51	0.33	0.23
90	9.18	2.30	1.02	0.57	0.37	0.26
100	10.20	2.55	1.13	0.64	0.41	0.28
150	15.30	3.83	1.70	0.96	0.61	0.43
200	20.40	5.10	2.27	1.28	0.82	0.57
250	25.50	6.38	2.83	1.59	1.02	0.71
300	30.60	7.65	3.40	1.91	1.22	0.85
400	40.80	10.20	4.53	2.55	1.63	1.13
500	51.00	12.75	5.67	3.19	2.04	1.42
600	61.27	15.30	6.80	3.83	2.45	1.70
700	71.40	17.85	7.93	4.46	2.86	1.98
800	81.60	20.40	9.07	5.10	3.26	2.27
900	91.80	22.95	10.20	5.74	3.67	2.55
1,000	102.00	25.50	11.33	6.38	4.08	2.83
1,500	153.00	38.25	17.00	9.56	6.12	4.25
2,000	204.00	51.00	22.67	12.75	8.16	5.67

To estimate Btuh

Btuh=500 x GPM x Δt

- Btuh = Btu/Hr
- GPM = Gallons per minute
- 500 = 8.33(Weight of one gallon of water) x 60 (minutes)
- Δt = Temperature difference F^0

One gpm will deliver approximately 10,000 Btuh with a 20 degree F delta T

Maximum flow through a boiler is boiler output divided by the temperature rise* divided by 500. **Please note most boilers require a 20-30 degree temperature rise.* **20^0F**

How many Btus @ various boiler temperature Delta T		
20^0F Delta T	**25^0F Delta T**	**30^0F Delta T**
3.45 GPM per Boiler HP	2.88 GPM per Boiler HP	2.30 GPM per Boiler HP
10,000 Btuh/GPM	12,500 Btuh/GPM	15,000 Btuh/GPM

	20^0F Rise	**25^0F Rise**	**30^0F Rise**
Btuh Output	**Gallons per minute**		
400,000	40	32	27
500,000	50	40	33
600,000	60	48	40
700,000	70	56	47
800,000	80	64	53
900,000	90	72	60
1,000,000	100	80	67
1,100,000	110	88	73
1,200,000	120	96	80
1,300,000	130	104	87
1,400,000	140	112	93
1,500,000	150	120	100
1,600,000	160	128	107
1,700,000	170	136	113
1,800,000	180	144	120
1,900,000	190	152	127
2,000,000	200	160	133
3,000,000	300	240	200
5,000,000	500	400	333
10,000,000	1,000	800	667

PIPING
Recommended Maximum Pipe Flow Rates based upon 20°F Delta T

Copper Pipe		
Pipe Size	Maximum Flow GPM	Btuh
½"	1 1/2	15,000
¾"	4	40,000
1"	8	80,000
1 ¼"	14	140,000
1 ½"	22	220,000
2"	45	450,000
2 ½"	85	850,000
3"	130	1,300,000

PEX Piping		
Pipe Size	Maximum Flow GPM	Btuh
3/8"	1.2	12,000
1/2"	2	20,000
5/8"	4	40,000
3/4"	6	60,000
1"	9.5	95,000

Steel Pipe		
Pipe Size	Maximum Flow GPM	Btuh
½"	2	15,000
¾"	4	40,000
1"	8	80,000
1 ¼"	16	160,000
1 ½"	25	250,000
2"	50	500,000
2 ½"	80	800,000
3"	140	1,400,000
4"	300	3,000,000
6"	850	8,500,000
8"	1,800	18,000,000
10"	3,200	32,000,000
12'	5,000	50,000,000

Sizing a Circulator

There are a couple short cuts to sizing a pump for a boiler. Most boilers are designed for a 20 to 30 degree rise through the boiler. To size a pump for a boiler and maintain a 20 degree rise through the boiler, divide the output of the boiler by 10,000 to get the proper GPM for a 20 degree rise.

If the boiler can handle a 30 degree rise, divide the output of the boiler by 15,000 to get the proper GPM.

Example: To see if the existing 40 GPM pump is large enough for the new boiler, let us look at the equipment. Our new boiler has a rated output of 800,000 with a design temperature rise of 20 degrees F. The existing pump is 40 GPM. If we divide the boiler output by 10,000, we see that the boiler will require an 80 GPM pump. This is double the GPM of the existing pump. Our flow would be half and the temperature rise would be double, possibly ruining the new boiler. In this case, we would have to replace the pump. If our new boiler can handle a 30 degree rise, we could divide it by 15,000 and find that the boiler will require a 53 GPM pump. The existing pump is too small for this boiler.

If you would like to see how we arrived at the 10,000 or 15,000 number, the following is the formula:

$$\text{GPM} = \frac{\text{Rated output of boiler}}{8.33 * 60 * \triangle\ ^\circ\text{F}}$$

or

$$\text{GPM} = \frac{\text{Rated output of boiler}}{500 * \triangle\ ^\circ\text{F}}$$

$$500 = 8.33 * 60$$

GPM = Gallons per minute flow rate

8.33 = Weight of a gallon of water

60 = Converts the formula from hours to minutes aka Gallons per Minute GPM

$\triangle$ °F Temperature rise through boiler is usually about 20-30 degrees F.

The following is the actual formula for a 20 degree rise for the 800,000 Btuh boiler:

$$\text{GPM} = \frac{800{,}000\ Btuh\ (Output\ of\ boiler)}{8.33*60*Temperature\ rise\ through\ boiler}$$

$$\text{GPM} = \frac{800{,}000\ Btuh\ (Output\ of\ boiler)}{500*20\ Degree\ rise}$$

$$80\ \text{GPM} = \frac{800{,}000\ Btuh\ (Output\ of\ boiler)}{10{,}000}$$

The following is the actual formula for a 30 degree rise for the 800,000 Btuh boiler:

$$\text{GPM} = \frac{800{,}000\ Btuh\ (Output\ of\ boiler)}{8.33*60*Temperature\ rise\ through\ boiler}$$

$$\text{GPM} = \frac{800{,}000\ Btuh\ (Output\ of\ boiler)}{500*30\ Degree\ rise}$$

$$53\ \text{GPM} = \frac{800{,}000\ Btuh\ (Output\ of\ boiler)}{15{,}000}$$

Calculate pump head

1. Measure longest run in feet. Include both supply and return.
2. Multiply by 1.5 to calculate fittings and valves
3. Multiply by 0.04 (4' head for each 100' of pipe ensures quiet operation)

For example, 100 feet is longest run
100 x 1.5 x .04 = 6 feet of head

Water

Water Density ᵛf

Temperature	**Density**		Specific Volume
⁰F	1 Lb/ft³	Lb/ gallon	**ᵛf**
32	63.41	8.48	0.01747
39	63.42	8.48	0.01602
50	63.40	8.48	0.01602
68	63.31	8.46	0.01605
86	63.15	8.44	0.01609
122	62.67	8.38	0.01621
140	62.35	8.34	0.01629
158	62.01	8.29	0.01638
176	61.63	8.24	0.01648
194	61.22	8.18	0.01659
212	60.78	8.13	0.01671

Typical Water Density	
One Cubic Foot Water =	62.43 lbs
	7.48052 gallons
	29.92 quarts
One pound of water =	27.72 cubic inches
One Gallon of Water =	0.1337 Cubic Feet

Water Capacity Steel Pipe			
Sch 40	**US Gallons per Foot**	**Sch 40**	**US Gallons per Foot**
Pipe Size Inches	**Water Capacity/ ft**	**Pipe Size Inches**	**Water Capacity/Ft**
½"	0.016	3"	0.390
¾"	0.023	4"	0.690
1"	0.040	5"	1.100
1 ¼"	0.063	6"	1.500
1 ½ "	0.102	8"	2.599
2"	0.170	10"	4.096
2 ½"	0.275	12"	5.815

Water Capacity Copper Tubing			
	US Gallons per Foot		
Pipe Size	**Type K**	**Type L**	**Type M**
3/8”	0.006	0.007	0.008
½”	0.011	0.012	0.013
5/8”	0.017	0.017	
¾”	0.023	0.025	0.027
1“	0.040	0.043	0.045
1 ¼”	0.063	0.065	0.068
1 ½”	0.089	0.092	0.095
2”	0.159	0.161	0.165
2 ½”	0.242	0.248	0.254
3”	0.345	0.354	0.363
4”	.608	.571	.634

Average System Water Content US Gallons	
Heating Element	Estimated Volume
Cast Iron Radiation	
Radiator, Large Tube	0.114 gal/ sq foot
Radiator, Thin Tube	0.056 gal/ sq foot
Convectors	1.5 Gal/10,000 Btu/Hr @ 200^{0}F
Baseboard	4.7 Gal/10,000 Btu/Hr @ 200^{0}F
Radiation Non Ferrous	
Convectors	0.64 Gal/10,000 Btu/Hr @ 200^{0}F
Baseboard ¾”	.37 Gal/10,000 Btu/Hr @ 200^{0}F
Fan Coil / Unit Htr.	.2 Gal/10,000 Btu/Hr @ 180^{0}F

Water Scalding Times

Temperature ^{0}F	***1st Degree***	***2nd/3rd Degree***
111.20	5 Hrs	7 Hrs
116.60	35 Mins	45 Mins
118.40	10Mins	14 Mins
122.00	1 Minute	5 Mins
131.00	5 Secs	22 Secs
140.00	2 Secs	5 Secs
149.00	1 Sec	2 Secs
158.00	1 Sec	1 Sec

Water Density			
Temperature	Temperature	Density	
^{0}F	^{0}C	Lb/ft^3	Lb/ gallon
32	0	63.41	8.48
39	3.89	63.42	8.48
50	10	63.40	8.48
68	20	63.31	8.46
86	30	63.15	8.44
122	50	62.67	8.38
140	60	62.35	8.34
158	70	62.01	8.29
176	80	61.63	8.24
194	90	61.22	8.18
212	100	60.78	8.13

Typical Water Calculations	
One Cubic Foot Water =	62.43 lbs
	7.48 gallons
	29.92 quarts
	1,728 Cubic Inches
One pound of water =	27.72 cubic inches @ 65 Degrees F
One Gallon of Water =	8.33 pounds
	0.1337 Cubic Feet
	4 Quarts
	8 Pints
	16 Cups

WATER

To estimate static pressure in system, multiply highest riser by 0.43 to get pressure at lowest point of system. Always add 4 pounds to get the right pressure for the building.
To estimate pump horsepower required - Horsepower = (GPM x Total head in feet) / 3960
To estimate flow rate of water through a pipe in gallons per minute - GPM = 0.0408 x $(\text{pipe diameter})^2$ x (water velocity)
To estimate weight of water in a given section of pipe in pounds- Lbs of water = 0.34 x pipe length(feet) x $(\text{pipe diameter})^2$
Maximum water velocity in pipes should be less than 6 feet per second @ 200^0F
To estimate flow rate of water through a pipe in gallons per minute = GPM = 0.0408 x $(\text{pipe diameter})^2$ x (water velocity)
To convert PSIG to feet of water, multiply PSIG x 2.307

Water Conversion Factors

US Gallons	X	8.34	=	Pounds
US Gallons	X	0.1338	=	Cubic Feet
US Gallons	X	231	=	Cubic Inches
US Gallons	X	3.7853	=	Liters
US Gallons	X	8.339	=	Pounds of Water
Cu Inch water	X	0.03613	=	Pounds
Cu Inch water	X	0.004329	=	US Gallons
Cu Inch water	X	0.576384	=	Ounces
Pounds Water	X	27.72	=	Cu Inches
Pounds Water	X	0.12	=	US Gallons
PSIG	X	2.307	=	Height of water in feet

Water Pressure to Feet Head			
Pounds Per Sq Inch	Feet Head	**Pounds Per Sq Inch**	Feet Head
1	2.31	**100**	230.90
2	4.62	**110**	253.98
3	6.93	**120**	277.07
4	9.24	**130**	300.16
5	11.54	**140**	323.25
6	13.85	**150**	346.34
7	16.16	**160**	369.43
8	18.47	**170**	392.52
9	20.78	**180**	415.61
10	23.09	**200**	461.78
15	34.63	**250**	577.24
20	46.18	**300**	692.69
25	57.72	**350**	808.13
30	69.27	**400**	922.58
40	92.36	**500**	1,154.48
50	115.45	**600**	1,385.39
60	138.54	**700**	1,616.3
70	161.63	**800**	1,847.2
80	184.72	**900**	2,078.1
90	207.81	**1,000**	2,309.00

Feet Head to Water Pressure			
Feet Head	Pounds Per Sq Inch	**Feet Head**	Pounds Per Sq Inch
1	.43	**100**	43.31
2	.87	**110**	47.64
3	1.30	**120**	51.97
4	1.73	**130**	56.30
5	2.17	**140**	60.63
6	2.60	**150**	64.96
7	3.03	**160**	69.29
8	3.46	**170**	73.63
9	3.90	**180**	77.96
10	4.33	**200**	86.62
15	6.50	**250**	108.27
20	8.66	**300**	129.93
25	10.83	**350**	151.58
30	12.99	**400**	173.24
40	17.32	**500**	216.55
50	21.65	**600**	259.85
60	25.99	**700**	303.16
70	30.32	**800**	346.47
80	34.65	**900**	389.78
90	38.98	**1,000**	433.00

Hot Water System Makeup:

Minimum connection size shall be 10% of largest system pipe or 1", whichever is greater 20" pipe should equal a 2" connection.

Sizing an Compression tank

Recommended Sizing for Compression tanks

Nominal Capacity Gallons	Sq ft Radiation
18	350
21	450
24	650
30	900
35	900
35	1100

A rule of thumb for sizing an compression tank is One gallon for each 23 square feet of radiation or One gallon for each 3,500 Btu of radiation. If you are going to size an compression tank, several of the manufacturers have on line calculators that will help you. If you still wish to do a manual calculation, here are the following calculations:

Compression Tank Sizing

Closed Tank	$Vt = Vs\frac{[(V2/V1) - 1] - 3\alpha\Delta t}{(Pa/P1) - (Pa/P2)}$
Diaphragm Tank	$Vt = Vs\frac{[(V2/V1) - 1] - 3\alpha\Delta t}{1 - (P1/P2)}$

Definitions:
Vt = Volume of compression tank in gallons
Vs = Volume of water in system in gallons
V1= Ground water temperature
V2= Design heating water temperature (180°F)
$\Delta T = T_2 - T_1$ °F
T_1 = Lower system temperature, typically 40-50°F at fill condition.
T_2 = Higher system design temperature, typically 180°- 220°F.
$P\alpha$ = Atmospheric pressure (14.7 Psia)
P_1 = System fill pressure Minimum System pressure (Psia)
P_2 = System operating pressure Maximum System pressure (Psia)
α = Linear Coefficient of expansion
Steel 6.5×10^{-6}
Copper 9.5×10^{-6}

When choosing a diaphragm tank, use the acceptance factor when choosing the size. The acceptance factor is the amount of space that is available in the tank.

To see how this formula works, let us look at a hypothetical building. Our building has a system volume of 2,000 gallons. The system will operate between 180°F and 220°F. The minimum pressure will be 10 pounds and the maximum pressure will be at 25 pounds. 14.7 has to be added to each pressure to get atmospheric pressure. For example, the low pressure will be 14.7 pounds plus 10 pounds = 24.7 pounds. The high pressure will be 14.7 pounds plus 25 pounds = 39.7 pounds. Our relief valve is set at 30 pounds. The system is steel pipe. The volume of the water is as follows:

$V_1 = 40^0F = 0.01602\ ft^2/lb$ (Ground water temperature)

$V_2 = 220^0F = 0.01677\ ft^2/lb$ (Design Temperature)

α = coefficient of thermal expansion for steel pipe is 6.5×10^{-6}

We will size a closed tank. Here is the formula again.

$$Vt = Vs\frac{[(^{V2}/_{V1}) - 1] - 3\alpha\Delta t}{(^{P\alpha}/_{P1}) - (^{P\alpha}/_{P2})}$$

$$Vt = 2000\frac{[(^{0.01677}/_{0.01602}) - 1] - 3(6.5\ x\ 0.000001)x\ 180)}{(^{14.7}/_{24.7}) - (^{14.7}/_{39.7})}$$

$$Vt = 2000\frac{0.0468 - 0.00351}{0.595 - 0.370}$$

$$Vt = 2000\frac{0.0433}{.225}$$

Vt = 385 Gallons

Estimated system water volume 35 Gallons per Boiler HP

Typical system fill pressure 10 Psi

Rule of Thumb for Compression tank Sizing

Steel Piping

Entering Pressure 10 pounds
Maximum Pressure 25 pounds
Entering Temperature 40^{0}F
Maximum Temperature 220^{0}F

Steel Piping		
System Capacity in Gallons	**Closed Compression tank**	**Diaphragm Tank**
200	39	23
300	58	34
400	77	46
500	96	57
600	116	69
700	135	80
800	154	92
900	173	103
1,000	193	115
1,100	212	126
1,200	231	138
1,300	250	149
1,400	270	160
1,500	289	172
1,600	308	183
1,700	327	195
1,800	347	206
1,900	366	218
2,000	385	229
2,500	481	287
3,000	578	344
3,500	674	401
4,000	770	458
4,500	867	516
5,000	963	573
6,000	1,156	688
7,000	1,348	802
8,000	1,541	917
9,000	1,733	1,032
10,000	1,926	1,146

Copper Piping

Based upon the following:
Entering Pressure 10 pounds
Maximum Pressure 25 pounds
Entering Temperature 40^{0}F
Maximum Temperature 220^{0}F

Copper Piping		
System Capacity in Gallons	**Closed Compression tank**	**Diaphragm Tank**
200	37	22
300	56	33
400	74	44
500	93	55
600	111	66
700	130	77
800	148	88
900	167	99
1,000	185	110
1,100	204	121
1,200	222	132
1,300	241	143
1,400	260	154
1,500	278	165
1,600	297	177
1,700	315	188
1,800	334	199
1,900	352	210
2,000	371	221
2,500	463	276
3,000	556	331
3,500	649	386
4,000	742	441
4,500	834	496
5,000	927	552
6,000	1,112	662
7,000	1,298	772
8,000	1,483	883
9,000	1,668	993
10,000	1,854	1,103

How to Estimate Hydronic System Volume

In many instances, you will need to calculate system water volume. This is useful when estimating water treatment and or glycol requirements. The following are some ideas that may help you to do that. The most accurate method is to measure and note the actual pipe sizes in the hydronic loop. This could be done by consulting the building blueprints. This is the most accurate method. There are several other rules of thumb that are used in the industry. I have a list of these below. A rule of thumb is that a hot water loop will be about 2/3 the size of a chilled water loop.

Rules of Thumb to Estimate System Volume

Multiply steel compression tank volume by 5.

35 – 50 gallons per Boiler HP

Pump GPM x 4

Compression tank volume is 20% of system volume

Rated tonnage of system x 10 gallons

The Salt Test

A common method for estimating system volume is to use salt because it is easy to test for, very soluble, and inexpensive. The disadvantage to this type of test is that the system has to be flushed at the end of the test, wasting water and chemicals. If not, the high chloride levels can be corrosive to the system.

Salt Test Procedure

Fill The system with fresh water. Circulate and flush the system until the water is clear. Eliminate all sources of water loss such as bleed, overflow, etc.

Measure the chloride Cl concentration in the system and estimate the system volume.

Add one pound of Table Salt (Sodium Chloride) per 1,000 gallons of estimated volume. This can be added in the pot feeder. Verify that the salt mixes thoroughly.

Allow one hour for the salt to be mixed into the system.

Re-Measure the chloride concentration

Multiply the estimated gallons of water by 76 ppm. Divide this by the difference (increase) in chloride concentration.

The answer will be the actual system volume

Example

Estimated Volume	1,000 Gallons
Initial Chloride Test	100 ppm
Final Chloride Test	180 ppm

Calculation

$$\frac{1{,}000\ gals.\times 76\ ppm}{(180-100)ppm} = 950 \text{ Actual Gallons in Loop}$$

**Flush system rapidly to return the chloride level to normal.*

Steam Heating Formulas & Rules of Thumb

Low Pressure Steam

Pipe Sizing to Assure Steam Velocities Below 15 FPS		
Pipe Size	**Lbs / Hr**	**Btu/ Hr**
2"	52.5	50,400
2 ½"	74.69	71,700
3"	115.47	110,850
4"	198.91	190,950
5"	312.34	299,850
6"	451.41	433,350
8"	799.22	767,250
10"	1,274.22	1,223,250
12"	1,793.75	1,722,000

Pipe Sizing to Assure Steam Velocities Below 40 FPS		
Pipe Size	**Lbs / Hr**	**Btu/ Hr**
2"	140	134,400
2 ½"	199	191,200
3"	307	295,600
4"	530	509,200
5"	833	799,600
6"	1,204	1,155,600
8"	2,131	2,046,000
10"	3,397	3,262,000
12"	4,783	4,592,000

Pipe Sizing to Assure Steam Velocities Below 50 FPS		
Pipe Size	**Lbs / Hr**	**Btu/ Hr**
2"	175.00	168,000
2 ½"	248.96	239,000
3"	384.90	369,500
4"	663.02	636,500
5"	1,041.15	999,500
6"	1,504.69	1,444,500
8"	2,664.06	2,557,500
10"	4,247.40	4,077,500
12"	5,979.17	5,740,000

Sizing a Dry Return Piping System				
Pipe Size	Btu/Hr		Pipe Size	Btu/Hr
1”	98,800		2 ½”	1,180,800
1 ¼”	208,320		3”	2,160,000
1 ½”	326,400		4”	4,636,800
2”	710,400			

Size Steam Main by Connected Radiation	
Radiation Sq Feet	Pipe Size Inches
75-125	1 ¼
125-175	1 ½
175-300	2
300-475	2 ½
475 – 700	3
700-1200	4
1200-1975	5
1975 – 2850	6

Boiler Feed Sizing Rules of Thumbs	
Pump GPM =	Evaporation Rate x 2
Tank Sizing =	Pump GPM x 20*
***Based on 20 minutes storage. One gallon storage per Boiler HP**	

Condensate Pump Sizing Rules of Thumb	
Pump GPM =	Evaporation Rate x 3
Tank Sizing =	Pump GPM x 1

Steam and Condensate Rules of Thumb	
Lbs. Steam / Hr =	Boiler Hp x 34.5
	Btuh divided by 960
Evaporation Rate in GPM =	Boiler Hp x .069
	Lbs. Steam Hr /500
Mbtu / Hr Output =	Boiler Hp x 33.4
EDR =	Boiler Hp x 139
	Btu/Hr /240
Evaporation Rate in GPM	EDR/1000 x 0.5
	Lbs/Steam/ Hr /500
Lb Steam condensate/ hr =	EDR divided by 4
Boiler Evaporation Rate =	Boiler HP x .069
	½ GPM per 240,000 Btuh
Boiler Feedwater Rate =	1 GPM per 240,000 Btuh Input
One cubic inch of water will produce One cubic foot of steam at atmospheric pressure.	
One cubic foot of steam exerts a mechanical force equal to that needed to lift 1,955 pounds one foot.	
Each nominal boiler horsepower requires 4-6 gallons of water per hour.	
At 5 steam pressure, the normal amount of flash steam should be 1.7% of the condensate volume.	

Miscellaneous Boiler Information

Closed Vessel Boiling Point @ PSI @ Sea Level	
Boiling Temperature	**Gauge Pressure**
212^{0}F	0 PSI
240^{0}F	10 PSI
259^{0}F	20 PSI
274^{0}F	30 PSI
287^{0}F	40 PSI
298^{0}F	50 PSI
316^{0}F	70 PSI
331^{0}F	90 PSI

Steam Flow through Orifices Discharging to Atmosphere			
	Steam Flow, Lb/Hr, when steam pressure is:		
Orifice Diameter	**2 PSI**	**5 PSI**	**10 PSI**
1/32"	.31	.47	.58
1/16"	1.25	1.86	2.3
3/32"	2.81	4.20	5.3
1/8"	4.5	7.5	9.4
5/32"	7.8	11.7	14.6
3/16"	11.2	16.7	21.0
7/32"	15.3	22.8	28.7
¼"	20.0	29.8	37.4
9/32"	25.2	37.8	47.4
5/16"	31.2	46.6	58.5
11/32"	37.7	56.4	70.7
3/8"	44.9	67.1	84.2
13/32"	52.7	78.8	98.8
7/16"	61.1	91.4	115
15/32"	70.2	105	131
½"	79.8	119	150

The following charts show the amount of steam lost per heating season with a leaking steam trap or a leaking pipe. To calculate costs of that leak, multiply the amount lost by the cost to generate 1,000 lbs of steam, typically $6.00 per 1,000 lbs. For example, if a system was running at 2psi and had 4,000 heating hours per winter with a leaking trap with a 1/8" hole, the system would lose about 18,000 pounds of steam per winter. The leaking trap would cost the owner about $108.00 per year.

Thousands of pounds of steam wasted through an orifice leak @ 2 PSI Steam Pressure						
Orifice Diameter	**Heating Hours per Year**					
	1000	2000	3000	4000	5000	6000
1/32"	0.31	0.62	0.93	1.24	1.55	1.86
1/16"	1.25	2.5	3.75	5	6.25	7.5
3/32"	2.81	5.62	8.43	11.24	14.05	16.86
1/8"	4.5	9	13.5	18	22.5	27
5/32"	7.8	15.6	23.4	31.2	39	46.8
3/16"	11.2	22.4	33.6	44.8	56	67.2
7/32"	15.3	30.6	45.9	61.2	76.5	91.8
¼"	20	40	60	80	100	120
9/32"	25.2	50.4	75.6	100.8	126	151.2
5/16"	31.2	62.4	93.6	124.8	156	187.2
11/32"	37.7	75.4	113.1	150.8	188.5	226.2
3/8"	44.9	89.8	134.7	179.6	224.5	269.4
13/32"	52.7	105.4	158.1	210.8	263.5	316.2
7/16"	61.1	122.2	183.3	244.4	305.5	366.6
15/32"	70.2	140.4	210.6	280.8	351	421.2
½"	79.8	159.6	239.4	319.2	399	478.8

Thousands of pounds of steam wasted through an orifice leak @ 5PSI Steam Pressure						
Orifice Diameter	**Heating Hours per Year**					
	1000	2000	3000	4000	5000	6000
1/32"	0.47	0.94	1.41	1.88	2.35	2.82
1/16"	1.86	3.72	5.58	7.44	9.3	11.16
3/32"	4.2	8.4	12.6	16.8	21	25.2
1/8"	7.5	15	22.5	30	37.5	45
5/32"	11.7	23.4	35.1	46.8	58.5	70.2
3/16"	16.7	33.4	50.1	66.8	83.5	100.2
7/32"	22.8	45.6	68.4	91.2	114	136.8
¼"	29.8	59.6	89.4	119.2	149	178.8
9/32"	37.8	75.6	113.4	151.2	189	226.8
5/16"	46.6	93.2	139.8	186.4	233	279.6
11/32"	56.4	112.8	169.2	225.6	282	338.4
3/8"	67.1	134.2	201.3	268.4	335.5	402.6
13/32"	78.8	157.6	236.4	315.2	394	472.8
7/16"	91.4	182.8	274.2	365.6	457	548.4
15/32"	105	210	315	420	525	630
½"	119	238	357	476	595	714

Thousands of pounds of steam wasted through an orifice leak @ 10 PSI Steam Pressure						
Orifice Diameter	**Heating Hours per Year**					
	1000	2000	3000	4000	5000	6000
1/32"	0.58	1.16	1.74	2.32	2.9	3.48
1/16"	2.3	4.6	6.9	9.2	11.5	13.8
3/32"	5.3	10.6	15.9	21.2	26.5	31.8
1/8"	9.4	18.8	28.2	37.6	47	56.4
5/32"	14.6	29.2	43.8	58.4	73	87.6
3/16"	21	42	63	84	105	126
7/32"	28.7	57.4	86.1	114.8	143.5	172.2
¼"	37.4	74.8	112.2	149.6	187	224.4
9/32"	47.4	94.8	142.2	189.6	237	284.4
5/16"	58.5	117	175.5	234	292.5	351
11/32"	70.7	141.4	212.1	282.8	353.5	424.2
3/8"	84.2	168.4	252.6	336.8	421	505.2
13/32"	98.8	197.6	296.4	395.2	494	592.8
7/16"	115	230	345	460	575	690
15/32"	131	262	393	524	655	786
½"	150	300	450	600	750	900

Boiler Pressure Designations	
Low Pressure	Steam up to 15 psig Hot Water Up to 160 psig & 250^0 F
Medium Pressure	Steam pressure up to 50 psig
High Pressure	Steam pressure greater than 50 psig

Boiler Ratings			
Boiler HP	**Btu/hr Output**	**Steam Lb/hr**	**EDR Rating Sq Feet**
20	670,000	690	2,790
30	1,005,000	1,035	4,185
40	1,340,000	1,380	5,580
50	1,675,000	1,725	6,975
60	2,010,000	2,070	8,370
70	2,345,000	2,415	9,765
80	2,680,000	2,760	11,160
100	3,350,000	3,450	13,950
125	4,185,000	4,313	14,438
150	5,025,000	5,175	20,935
200	6,695,000	6,900	27,915
250	8,370,000	8,625	34,895
300	10,045,000	10,350	41,875
350	11,720,000	12,075	48,825
400	13,390,000	13,800	55,920
450	15,064,000	15,520	63,000
500	16,740,000	17,250	69,790
600	20,085,000	20,700	83,750
650	21,759,000	22,425	91,000
700	23,432,000	24,150	98,000
750	25,106,000	25,875	105,000
800	26,780,000	27,600	112,000
1000	33,475,000	34,500	140,000

Boiling Temperature of Water at Different Altitudes

Boiling Temperature of Water at Various Altitudes and Pressures					
	Gage Pressure				
Altitude Feet	**0**	**1**	**5**	**10**	**15**
-500	212.8	216.1	227.7	239.9	250.2
-100	212.3	215.5	227.2	239.4	249.9
Sea Level	212.0	215.3	227.0	239.3	249.7
500	211.0	214.4	226.3	238.7	249.2
1,000	210.1	213.5	225.5	238.1	248.6
1,500	209.4	212.7	225.0	237.6	248.2
2,000	208.2	211.7	224.1	236.8	247.7
2,500	207.3	210.9	223.4	236.3	247.2
3,000	206.4	210.1	222.7	235.7	246.7
3,500	205.5	209.2	222.1	235.1	246.2
4,000	204.7	208.4	221.4	234.6	245.7
4,500	203.7	207.5	220.7	234.0	245.2
5,000	202.7	206.8	218.7	233.4	244.7
6,000	200.9	205.0	217.3	232.4	243.8
7,000	199.1	203.3	216.1	231.3	242.9
8,000	197.4	201.6	214.8	230.3	242.0
9,000	195.7	200.0	213.5	229.3	241.3
10,000	194.0	198.4	212.3	228.3	240.4

Properties of Saturated Steam					
Gage Pressure	Temp F	Volume of 1lb steam cu.ft.	Sensible Heat	Latent Heat	Total Btu
0	212.0	26.8	180	970	1150
1	215.5	25.2	183	968	1151
2	218.7	23.5	187	966	1153
3	221.7	22.3	190	964	1154
4	224.5	21.4	192	962	1154
5	227.3	20.1	195	960	1155
6	229.9	19.4	198	959	1157
7	232.4	18.7	200	957	1157
8	234.9	18.4	201	956	1157
9	237.2	17.1	205	954	1159
10	239.5	16.5	207	953	1160
11	241.7	16	209	951	1161
12	243.8	15.3	212	949	1163
13	245.9	15	214	948	1164
14	247.9	14.3	216	947	1165
15	249.8	14	218	945	1163

Steam Capacity per Boiler HP				
	Pounds of dry saturated steam per boiler HP @ system pressure (PSIG) at a given feed water temperature			
	Steam Pressure			
Feed Water Temperature	0	2	10	15
100	30.9	30.8	30.6	30.6
110	31.2	31.2	30.9	30.8
120	31.5	31.4	31.2	31.1
130	31.8	31.7	31.5	31.4
140	32.1	32	31.8	31.7
150	32.4	32.4	32.1	32.0
160	32.7	32.7	32.4	32.4
170	33.0	33.0	32.7	32.6
180	33.4	33.3	33.0	32.9
190	33.8	33.7	33.4	33.3
200	34.1	34.0	33.7	33.6
212	34.5	34.4	34.2	34.1
220	34.8	34.7	34.4	34.3
227	35.0	34.9	34.7	34.5
230	35.2	35.0	34.8	34.7

Typical Water Quality Parameters in Steam Systems		
Feedwater	Softness	Less than 1 ppm
Boiler Water	Hardness	Less than 1 ppm
Boiler Water	Ph	7-9 or 9.5-11*
Boiler Water	TDS	1,500-3000 ppm 2,000-4000 microsiemens
Boiler Water	Sulfite	30-60 ppm
Boiler Water	Hydroxyl Alkalinity	200 - 400 ppm
Condensate	pH	8.2-9.0
Condensate	TDS	Less than 20 ppm
***Verify with boiler manufacturer**		

Common Heating Abbreviations

Symbols		
Δt	Delta T or Temperature Difference	
#	Pound	
"	Inches	
'	Foot	
	A	
AC	Alternating current	
AHU	Air handling unit	
AMB	Ambient	
	B	
BTU	British Thermal Unit	
BTUH	British Thermal Unit per Hour	
	C	
Cc	Cubic centimeter	
CFM	Cubic feet per minute	
CFS	Cubic feet per second	
CI	Cast Iron	
CL	Center Line	
CPVC	Chlorinated poly vinyl chloride	
Cu ft	Cubic foot	
Cu In	Cubic Inches	
D		
DC	Direct current	
DD	Degree day	
DEG	Degree	
Deg F or ^{0}F	Degree Fahrenheit	
Deg C or ^{0}C	Degree Celsius	
DHW	Domestic Hot water	
Diam	Diameter	
E		
EAT	Entering air temperature	
	F	
Fp	Freezing Point	
Fpm	Feet per minute	
Fps	Feet per second	
Ft	Foot	

G	
GPH	Gallons per hour
GPM	Gallons per minute
	H
HDD	Heating degree days
Hg	Mercury
HHWP	Heating hot water pump
HHWR	Heating hot water return
HHWS	Heating hot water supply
HPS	High pressure steam
Hs	Sensible heat
HVAC	Heating Ventilating & Air Conditioning
HWP	Hot water pump
HWR	Hot water return
HWS	Hot water supply
Hx	Heat exchanger
	I
In	Inches
	K
Kwh	Kilowatt per Hour
	L
LAT	Leaving air temperature
Lb	Pound
LH	Latent heat
LL	Low Limit
LPS	Low pressure steam
LRA	Locked rotor amps
LWCO	Low water cutoff
	M
MPT	Male pipe thread
	N
NC	Normally closed
NO	Normally open
	O
Oz	Ounce
	P
PRV	Pressure reducing valve
Psi	Pounds per square inch

P			T	
Psia	Pounds per square inch, absolute		TDH	Total dynamic head
Psig	Pounds per square inch, gauge		TEMP	Temperature
PVC	Poly vinyl chloride		TH	Thermometer
R			V	
Rpm	Revolutions per minute		V	Volt
Rps	Revolutions per second		W	
RV	Relief valve		W	Watt
S			WC	Water column
Sec	Second		Whr	Watt Hour
Sp gr	Specific gravity		Wmin	Watt minute
Sp ht	Specific heat			
Sq ft	Square foot			
Sq in	Square inch			

Degree Day Definition

The daily mean temperature is obtained by adding together the maximum and minimum temperatures reported for the day and dividing the total by two. Each degree of mean temperature below 65 is counted as one heating degree-day. Thus, if the maximum temperature is 70^0F and minimum 52^0F, four heating degree-days would be produced. (70 + 52 = 122; 122 divided by 2 = 61; 65-61 = 4.) If the daily mean temperature is 65 degrees or higher, the heating degree-day total is zero. The following Heating Design Temperatures are based on the 97 ½% temperatures for the locale. The area will be colder than the design temperature for 2 ½% of the winter

Degree Days and Design Temperatures

ST	Station	Heating Degree Days	Heating Design Temp F
AL	Birmingham	2551	21
	Huntsville	3,070	16
	Mobile	1,560	29
	Montgomery	2,291	25
AK	Anchorage	10,864	-18
	Fairbanks	14,279	-47
	Juneau	9,075	1
	Nome	14,171	-27
AZ	Flagstaff	7,152	4
	Phoenix	1,765	34
	Tucson	1,800	32
	Yuma	974	39
AR	Fort Smith	3,292	17
	Little Rock	3,219	20
	Texarkana	2,533	23
CA	Fresno	2,611	30
	Long Beach	1,803	43
	Los Angeles	2,061	43
	Los Angeles	1,349	40
	Oakland	2,870	36
	Sacramento	2,502	32
	San Diego	1,458	44
	San Francisco	3,015	38
	San Francisco	3,001	40
CO	Alamosa	8,529	-16
CO	Colorado Springs	6,423	2
	Denver	6,283	1
	Grand Junction	5,641	7
CO	Pueblo	5,462	0
CT	Bridgeport	5,617	9
	Hartford	6,235	7
	New Haven	5,897	7
DE	Wilmington	4,930	14
DC	Washington	4,224	17
FL	Daytona		
	Fort Myers	442	44
	Jacksonville	1,239	32
	Key West	108	57
	Miami	214	47
	Orlando	766	38
	Pensacola	1,463	29
	Tallahassee	1,485	30
	Tampa	683	40
	West Palm Beach	253	45
GA	Athens	2,929	22
	Atlanta	2,961	22
	Augusta	2,397	23
	Columbus	2,383	24
	Macon	2,136	25
	Rome	3,326	22
	Savannah	1,819	27
HI	Hilo	0	62
HI	Honolulu	0	63
ID	Boise	5,809	10
	Lewiston	5,542	6
	Pocatello	7,033	-1
IL	Chicago (Midway)	6,155	0
	Chicago (O'Hare)	6,639	-4
	Chicago	5,882	2
	Moline	6,408	-4
IL	Peoria	6,025	-4
	Rockford	6,830	-4
	Springfield	5,429	2
IN	Evansville	4,435	9
	Fort Wayne	6,205	1
	Indianapolis	5,699	2
	South Bend	6,439	1
IA	Burlington	6,114	-3
	Des Moines	6,588	-5
	Dubuque	7,376	-7
	Sioux City	6,951	-7
	Waterloo	7,320	-10
KS	Dodge City	4,986	5
	Goodland	6,141	0
	Topeka	5,182	4
	Wichita	4,620	7

KY	Covington	5,265	6
	Lexington	4,683	8
	Louisville	4,660	10
LA	Alexandria	1,921	27
	Baton Rouge	1,560	29
	Lake Charles	1,459	31
	New Orleans	1,385	33
	Shreveport	2,184	25
ME	Caribou	9,767	-13
	Portland	7,511	-1
MD	Baltimore	4,654	13
	Baltimore	4,111	17
	Frederick	5,087	12
MA	Boston	5,634	9
	Pittsfield	7,578	-3
	Worcester	6,969	4
MI	Alpena	8,506	-6
MI	Detroit (city)	6,232	6
	Escanaba	8,481	-7
	Flint	7,377	1
	Grand Rapids	6,894	5
	Lansing	6,909	1
	Marquette	8,393	-8
	Muskegon	6,696	6
	Sault Ste. Marie	9,048	-8
MN	Duluth	10,000	-16
MN	Minneapolis	8,382	-12
	Rochester	8,295	-12
MS	Jackson	2,239	25
	Meridian	2,289	23
	Vicksburg	2,041	26
MO	Columbia	5,046	4
	Kansas City	4,711	6
	St. Joseph	5,484	2
	St. Louis	4,900	6
	St. Louis	4,484	8
	Springfield	4,900	9
MT	Billings	7,049	-10
	Great Falls	7,750	-15
	Helena	8,129	-16
	Missoula	8,125	-6
NE	Grand Island	6,530	-3
	Lincoln	5,864	-2
	Norfolk	6,979	-4
	North Platte	6,684	-4
	Omaha	6,612	-3
	Scottsbluff	6,673	-3
NV	Elko	7,433	-2
	Ely	7,733	-4
	Las Vegas	2,709	28
	Reno	6,332	10
	Winnemucca	6,761	3
NH	Concord	7,383	-3
NJ	Atlantic City	4,812	13
	Newark	4,589	14
	Trenton	4,980	14
NM	Albuquerque	4,348	16
	Raton	6,228	1
	Roswell	3,793	18
NM	Silver City	3,705	10
NY	Albany	6,875	-1
	Albany	6,201	1
	Binghamton	7,286	1
	Buffalo	7,062	6
	NY(central park)	4,871	15
	NY(Kennedy	5,219	15
	NY (LaGuardia)	4,811	15
	Rochester	6,748	5
NY	Schenectady	6,650	1
	Syracuse	6,756	2
NC	Charlotte	3,181	22
	Greensboro	3,805	18
	Raleigh	3,393	20
	Winston-Salem	3,595	20
ND	Bismarck	8,851	-19
	Devils Lake	9,901	-21
	Fargo	9,226	18

	Williston	9,243	-21
OH	Akron-Canton	6,037	6
	Cincinnati	4,410	6
	Cleveland	6,351	5
	Columbus	5,660	5
	Dayton	5,622	4
	Mansfield	6,403	5
	Sandusky	5,796	6
	Toledo	6,494	1
	Youngstown	6,417	4
OK	Oklahoma City	3,725	13
	Tulsa	3,860	13
OR	Eugene	4,726	22
	Medford	5,008	23
	Portland	4,635	23
	Portland	4,109	24
	Salem	4,754	23
PA	Allentown	5,810	9
	Erie	6,451	9
	Harrisburg	5,251	11
	Philadelphia	5,144	14
	Pittsburgh	5,987	5
PA	Pittsburgh	5,053	7
	Reading	4,945	13
	Scranton	6,254	5
	Williamsport	5,934	7
RI	Providence	5,954	9
SC	Charleston	2,033	27
	Charleston	1,794	28
	Columbia	2,484	24
SD	Huron	8,223	-14
	Rapid City	7,345	7
SD	Sioux Falls	7,839	-11
TN	Bristol	4,143	14
	Chattanooga	3,254	18
	Knoxville	3,494	19
	Memphis	3,232	18
	Nashville	3,578	14
TX	Abilene	2,624	20
	Austin	1,711	28
	Dallas	2,363	22
	El Paso	2,700	24
	Houston	1,396	32
	Midland	2,591	21
	San Angelo	2,255	22
	San Antonio	1,546	30
	Waco	2,030	26
	Wichita Falls	2,832	18
UT	Salt Lake City	6,052	8
VT	Burlington	8,269	-7
VA	Lynchburg	4,166	16
	Norfolk	3,421	22
	Richmond	3,865	17
	Roanoke	4,150	16
WA	Olympia	5,236	22
	Seattle-Tacoma	5,145	26
	Seattle	4,424	27
	Spokane	6,655	2
WV	Charleston	4,476	11
	Elkins	5,675	6
	Huntington	4,446	10
	Parkersburg	4,754	11
WI	Green Bay	8,029	-9
	La Crosse	7,589	-9
	Madison	7,863	-7
WI	Milwaukee	7,635	-4
WY	Casper	7,410	-5
	Cheyenne	7,381	-1
	Lander	7,870	-11
	Sheridan	7,680	8

Definitions

Air Change: the amount of air that is required to completely replace the air in the boiler and associated flue passages.

Air, Primary: Air that mixes with the fuel to provide combustion.

Air, Secondary: Air that mixes with the flue gases to provide proper turbulence to allow complete combustion.

Air Separator: A device located in the supply pipe for a hydronic boiler that removes the entrained air from the water.

Air shutter: A device that controls the airflow to the burner

Air, Tertiary: Air from the boiler room that is introduced to the flue to overcome excessive chimney draft. It is sometimes called Dilution Air.

Backflow Preventer: A device that will limit the backflow of boiler water into the potable water in a building or system.

Barometric Damper: A damper that is installed in the flue piping that will control the excessive draft in a category1 type boiler by introducing boiler room air.

Boiler: A closed vessel that heats water or creates steam

Boiler Design Temperature: It is the outside temperature at which the heating system can still provide heat to the building. It will be one of the coldest temperatures during an average winter.

Boiler, High Pressure: A boiler, which generates steam to pressures above 15 Psig

Boiler, Low Pressure: A boiler, which generates steam to pressures below 15 Psig

Boiler, Hydronic: A boiler, which heats water below the flash point.

Boiler, Cast Iron: A boiler, which uses cast iron as its heat exchanger.

Boiler, Copper: A boiler, which uses copper as its heat exchanger.

Boiler, Steel: A boiler, which uses steel as its heat exchanger.

Boiler, Fire Tube: A steel boiler where the flue gases travel through the tubes inside the boiler.

Boiler, Water Tube: A steel boiler where the flue gases travel around the tubes inside the boiler.

Boiler, Modular: A heating system consisting of several smaller boilers.

Breeching: A conduit that transports the combustion by products from the boiler to the outside or to the chimney. It is also called a flue.

Btu (British Thermal Unit): The amount of heat required to raise one pound of water, one degree F

Btuh: Btu's in one hour

Burner: A mechanical device that mixes air and fuel to provide ignition and combustion of the fuel.

Burner, Atmospheric: A burner that uses natural draft and gas pressure to provide combustion.

Burner, Power: A burner that uses an internal blower to mix the fuel and the air for combustion.

Carbon Dioxide: This is a gas that is produced as a by-product of combustion. It is also referred to as CO2.

Carbon Monoxide: This deadly gas is odorless and tasteless. It is produced when there the combustion is out of adjustment. It is often referred to as CO.

Compression Tank: A tank that is used in a hydronic system that will absorb the expansion of the water, sometimes called an Compression tank.

Combustion Air: The air that is introduced from the outside that is required for the proper combustion of the fuel.

Combustion Analyzer: A device that measures the flue gas from a boiler and displays the different components. It will also display the efficiency of the boiler.

Condensate: Condensed water because of the removal of latent heat from a gas.

Condensing Boiler: A boiler that is designed to allow the flue temperatures to drop below the dew point temperature.

Control, Operating: A device that starts or stops the burner. This is usually set for a lower temperature or pressure than the Limit Control.

Control, Limit: A device that starts or stops the burner. This is usually set for a higher pressure or temperature than the operating control. In most applications, it has a manual reset feature.

Dew Point Temperature: The temperature at which warm humid air is cooled enough to allow the water vapor to condense into water.

Dirt Leg: A series of nipples and a pipe cap that are installed just before the train to capture any dirt that is in the gas line before it enters the gas train.

Draft: The pressure differential between atmospheric pressure and the pressure in the flue and boiler.

Draft Diverter: An air opening that introduces tertiary air to the flue after the main combustion.

Draft, Mechanical: The pressure differential between atmospheric pressure and the pressure in the flue and boiler that is induced because of a fan or blower.

Draft, Natural: The pressure differential between atmospheric pressure and the pressure in the flue and boiler without a fan or blower.

Dual Fuel Burner: A burner, which has two fuel sources that it can use. It is usually natural gas and #2 fuel oil.

Emergency Door Switch: This manual switch is located at all exits from a boiler room that will shut off the boiler in the event that it is engaged.

Expansion Tank: A tank that is used in a hydronic system that will absorb the expansion of the water once it is heated. It is sometimes called a Compression Tank.

Firing Rate: The burning rate of fuel and air in the burner.

Firing Rate Control: A control that senses the temperature or pressure of the heating system. It will regulate the burner between low and high fire to meet the desired set point. It is sometimes called the modulating control.

Flue: A conduit that transports the combustion by products from the boiler to the outside or to the chimney. It is also called a breeching.

Flue Gases: These byproducts of combustion are produced by the burner. They will be vented from the boiler with a flue.

Fuel Train: A series of components, including gas pressure regulator and gas valves, that are located in the gas piping directly attached to the burner. This is also called a gas train.

Gas Pressure Regulator: A device that controls the gas pressure supplied to the burner.

Gas Pressure Switch: A safety device that senses the available gas pressure and will shut the boiler off in the event that the pressure is outside of the setting. There are usually two types of gas pressure switches on a boiler. The High Gas Pressure Switch that is located in the gas train downstream of the gas pressure regulator and the electric gas valves. It will shut the boiler off if the gas pressure is higher than the setting. The Low Gas Pressure Switch is located downstream of the main gas pressure regulator. It will shut the boiler off if the gas pressure is below the setting.

Heat, Latent: The amount of heat required to cause a change of state.

Heat, Sensible: The amount of heat required to cause a change in temperature

Heating Medium: The material that the boiler heats. It could be steam, water or some other type of fluid.

High Fire: This is the highest design firing rate of the burner. It is the 100% firing rate.

Hydronic System: A heating system that uses water as the heating medium instead of steam.

Lag Boiler: The boiler that is not the first boiler to start when there is a call for heat.

Lead Boiler: The boiler that is the first boiler to start on a call for heat.

Life Cycle Cost: This is the amount of money that the system costs the owner over the estimated life of the unit. It will include fuel and repair costs as well as estimated repair parts.

Lockout: A safety shutdown that requires a manual reset of the control or safety device.

Low Fire: This is the lowest design firing rate of the burner.

Low Fire Start: This switch verifies that the burner is in the "Low Fire" position before opening fuel valves.

Low High Low Fire: A burner that starts at low fire and then goes to high fire if there is still a call for heat. As the temperature or pressure gets close to the set point on the firing rate control, the burner will drop to low fire.

Low High Off Fire: A burner that starts at low fire and then goes to high fire if there is still a call for heat. The burner will stay at high fire until the call for heat has ceased.

Low Water Cutoff: A device that senses the water level inside the boiler and will shut down the burner if the water level drops to an unsafe level.

Modulating Burner: A burner that will operate at any position from low to high fire to meet the demands of the modulating control.

Modulating Control: A control that senses the heating medium and will send a signal to the burner that will set the burner at any position from low to high fire.

Non-Condensing Boiler: A boiler that is designed to keep the flue temperatures above the dew point temperature.

Pilot, Continuous: It is a pilot flame that burns all the time, regardless of whether the burner is firing.

Pilot, Intermittent. It is a pilot that lights when there is a call for heat. The pilot will stay light during the entire time that the main burner is firing.

Pilot, Interrupted: It is a pilot that lights when there is a call for heat. The pilot will shut off once the main flame is established.

Pot Feeder: A device that is used to introduce water treatment into a heating system.

PPM: Part per Million.

Prepurge: On a call for heat, the burner blower starts to purge the boiler combustion chamber and flue passages of any unburnt fuels. It will operate for a duration long enough to provide several air changes inside the boiler.

Proprietary Parts: parts that are only available from the manufacturer or have limited distribution.

Post purge: The burner blower will operate for a time after the call for heat has been satisfied to purge any unburnt fuel.

Relief Valve: A valve located on a boiler that will relieve the internal boiler pressure if the pressure rises to the rating of the relief valve.

Reset Control: A control that will lower the supply temperature in a hydronic system as the outside temperature increases.

Sidewall Venting: Boiler flue that is piped to the sidewall of the building rather than a chimney or stack.

Siphon A water seal used on steam boilers that protects the inner workings of the steam pressure control from steam.

Spill Switch: A device that is located by a draft diverter or a barometric damper that senses rollout of the flue gases and shuts off the burner.

Index

A-B

C-D

E-F

G-H

I-J

M-N

O-P

Q-R

S-T

U-V

U-Z

Made in United States
North Haven, CT
12 November 2024